AF598481

Improving Productivity by CLASSIFICATION, CODING, AND DATA BASE STANDARDIZATION

MANUFACTURING ENGINEERING AND MATERIALS PROCESSING

A Series of Reference Books and Textbooks

SERIES EDITORS

Geoffrey Boothroyd
Department of Mechanical Engineering
University of Massachusetts
Amherst, Massachusetts

George E. Dieter
Dean, College of Engineering
University of Maryland
College Park, Maryland

1. Computers in Manufacturing, *U. Rembold, M. Seth, and J. S. Weinstein*
2. Cold Rolling of Steel, *William L. Roberts*
3. Strengthening of Ceramics: Treatments, Tests, and Design Applications, *Henry P. Kirchner*
4. Metal Forming: The Application of Limit Analysis, *Betzalel Avitzur*
5. Improving Productivity by Classification, Coding, and Data Base Standardization: The Key to Maximizing CAD/CAM and Group Technology, *William F. Hyde*

OTHER VOLUMES IN PREPARATION

Improving Productivity by CLASSIFICATION, CODING, AND DATA BASE STANDARDIZATION

The Key to Maximizing CAD/CAM and Group Technology

WILLIAM F. HYDE
Brisch, Birn & Partners
Fort Lauderdale, Florida

MARCEL DEKKER, INC. New York and Basel

Library of Congress Cataloging in Publication Data

Hyde, William F., [date]
Improving productivity by classification, coding, and data base standardization.

(Manufacturing engineering and materials processing ; 5)
Includes index.
1. Industrial productivity. 2. Group technology. 3. Efficiency, Industrial. I. Title. II. Series.
T58.8.H92   658.5'1   81-7821
ISBN 8247-1404-0   AACR2

MARCEL DEKKER, INC.

270 Madison Avenue, New York, New York 10016

Current printing (last digit):
10 9 8 7 6 5 4 3 2 1

PRINTED IN THE UNITED STATES OF AMERICA

To my wife Carol

Foreword

One could say "it" all started over 30 years ago in London, England, and it is therefore fitting that someone from the United Kingdom should be asked to write the Foreword to this unusual book. I have the honor to do so for two reasons: first, it has been my privilege to know and work with Messrs. Gombinski and Hyde for so many years; and second, it has also been my privilege to have helped pioneer the use of the techniques referenced in this book.

So many designs are made, so much money wasted, so many people hurt because so-called entrepreneurs and business managers have failed all their lives to recognize one simple rule that is taught to first-line managers—*"Get the facts!"* No decision can be taken with confidence when the facts relevant to the decision are denied to the manager.

The author, like many others in the developed industrialized world, is concerned about productivity. However, he is particularly concerned about productivity in the United States! He used this concern to point the way to some of the causes of reduced productivity and some of the cures. Although he references problems experienced in America, they are not unique and are common to every country where managers must take or arrive at a decision.

Regarding other basics, he also repeats a favorite expression of mine and many others—"fear no longer motivates." The worker today is insulated and, in good measure, protected by social programs that ameliorate the fears of loss of job. In the present world, it is the job challenge and job design which matter so very, very much. Money is not a true motivator. After basic needs have been met, money, for a variety of reasons (not the least of which is confiscatory taxation rates), fails as a motivator by itself. Job satisfaction must also be present.

In his approach to the subject matter of this book, the author goes back 100 years to Dewey, in 1876, and also to F. W. Taylor, who, in the early years of management, strongly advocated classification and coding as a means to manage the many facets of a business and to prevent waste. In my business life, I have written many memoranda headed "Management Waste." Management waste takes place to a greater degree than is realized in all industrialized countries. For example, the cost of creating a new component part for production in the United States is something on the order of $2000 each. An analysis reveals that up to 46% of what it requires to create and prepare designed parts has been found to be redundant.

Classification and coding when applied correctly has been known to show a rate of investment return of from 75% to over 400% per annum. This is not surprising when it can be shown that there are in existence 109 different names for the one same component. Designers and draftsmen believe that they are paid to design, but are they designing when they duplicate that which has been done once, twice, or even thrice before?

Much has been written about group technology, and 95% of it has been nonsense. The Russians led the way, but people in the United Kingdom and the United States rushed into something which they did not understand, but about which they wished to appear knowledgeable—because the media were writing about it. The Russians in the 1930s approached the problem out of need. They produced a realistic and pragmatic solution to their problems. They had limited financial resources coupled with low volume, long lead times, and horrendous work-in-progress investment—in short, grossly inefficient management. Thus, out of need was developed modular technological manufacturing.

That job enrichment has resulted from true group technology is because team spirit emerges. Working in a group is now becoming the "in" thing, but is this not what the Japanese have been doing for years and years, and is this the concept of GT that most American managers have?—I am afraid not, from personal observations. The development of the use of computers has greatly enhanced the value of classification and coding as a tool of management. Savings in costs, using the computer and classification and coding together, can be astronomical. This has been my own experience. The author has made a case for managers to take an in-depth look at classification and coding of the data vital to improving the capabilities in decision making. I heartily endorse his case with three very profitable firsthand experiences.

I am sure that this book will become a standard textbook in many universities, polytechnical institutions, and colleges throughout the world and will be a valuable contribution to the practice of management.

William Jack
Bridge of Allan
Stirling, Scotland

Preface

Why a book on industrial classification, coding, standardization, and group technology and CAD/CAM? For a variety of reasons, two of the most important being

1. There is an information vacuum. Several books have been written on group technology in which classification and coding were touched upon. None have been written emphasizing the prerequisite role of classification and coding for effective group technology application. None have been written that treat the management of data as a continuing source of profit.
2. Despite, or perhaps because of, a spate of business periodical articles on these subjects, a great deal of confusion and misinformation is threatening to damage the reputation of these time-proven management tools.

Now CAD/CAM has burst upon the scene and without variety controls may well harm rather than help our economy.

For nearly 16 years, I have lectured professionally on these subjects. My colleagues and I, in addition, have addressed an endless number of meetings sponsored by our various professional societies. We have contributed many articles to the business press and to periodical publishers. We have hosted six international conferences. Finally, we have visited and presented these subjects to several hundred institutions and business firms in this country alone.

For the past several years, I have polled my audiences for their respective definitions of these subjects, especially group technology. Three out of four either admitted they didn't know or didn't know but thought they did—which is even worse.

This book is an effort to try to dispel some of the myths and misinformation spread by well-meaning but less-than-informed people. Specifically, the book attempts to

Clearly define all terms
Provide some theory
Present some practical examples

The objective is to educate managers to appreciate at least some of the "do's" and "don'ts" we have experienced as a firm. Thirty-one years have come and gone since Joseph Gombinski took an idea by Edward Brisch and made it an effective management tool to control data. Twenty years have elapsed since Gombinski and his colleagues adapted the Russian Mitrofanov's group technology for Western use in a variety of applications. It does seem time to write about it.

There are those whose contributions to this book I wish to acknowledge:

My beloved wife, Carol Whipple Hyde, for her superb editing of the text, encouragement, and great support.

Mrs. Helen Simard, for her patient effort in typing and preparing the manuscript.

William Jack, for his commentary and his testimony on how he managed better using classified and coded data and group technology.

Joseph Gombinski, who made classification, coding, and group technology into an effective management tool.

Serge A. Birn (deceased), who said: "Now that we have it [Brisch], you run it"—and left me to do it.

Many, many thanks.

William F. Hyde

Contents

Improving Productivity by CLASSIFICATION, CODING, AND DATA BASE STANDARDIZATION

1

Introduction to the Data Management Problem

1.0 DATA, FACTS, AND DECISIONS

Managing is decision making, and the first commandment of decision making according to the gospel as written by Taylor, Gilbreth, et al. is: *Get the facts!*

Managers' decisions are usually made either (1) to resolve an existing problem or (2) to prevent one from developing. We can classify these kinds of decisions and code them. Class 1 decisions are tactical, and class 2 decisions, the problem preventers, are strategic. Regardless of which class of decision is to be made, facts (i.e., information that is reliable and relevant) are required.

Data are real or assumed premises that can be given as an argument or an inference. Reliable, accurate, and relevant data are facts. Sufficient available facts are the keystone for sound decision making.

Because data can be the result of inference or assumption, they are not necessarily facts. This is the real distinction between data and facts. Data are not necessarily truths; facts always are.

For example, in a recent assignment, investigation of data in a computer file for inventory control showed considerable disparity in the quantities on hand in the file versus those in the stockroom. The computer data file was not properly maintained and the integrity of the system was compromised. Data were not facts! Another example where data are not facts is the following:

The data items, in this case, are angular contact bearings. The file contained stock numbers at variance with the stockroom. An MRC5206-SBKG was shown stocked under 022-05006-00. The actual number was 002-65106-00.

Yet another example found was the same datum identified by key word for retrieval duplicated because of a different key word. This fault created 132 sets of duplicates and triplicates, and one quadruplicate. Imagine the quality of decisions on inventory levels, order quantities, usage, price structure, and the like on such unreliable data.

There was a time when management was handicapped by insufficient data. The manager collected the meager data available, analyzed the data, synthesized facts from the data, such as they were, and then made a calculated, intuitive guess.

Then came Taylor, who said that decision making without all the necessary facts was not very scientific. He said, using the example of work planning, that using time estimates based on past work performed without a record of *how* it was done was not reliable. He said that the practice of management was an art when it should be a science.

Whether you think Taylorism is dead or not is unimportant. Some of what he had to say is very relevant. Taylor's postulation was that to manage requires a capability to control. To control requires some form of factual measure. The facts of the measure must be comprised of sufficient data that can be proven by reproduction (i.e., to predict the end result). Now that is scientific, wouldn't you say?

The problem we have with data is the same today as it was then—how to distinguish fact from fiction. But today, the problem rarely involves a shortage of data. We have machinery to manipulate a vast amount of data in a very short period of time. However, to test them for integrity before accepting them as fact is just as important. Otherwise, the machinery is just helping managers make poor decisions more quickly.

Data can be tested if they can be retrieved. The real trick is to be able to make the data visible, so that we can find the relevant data when we need them. This has not been done too well in this country until recently.

Some years ago one of the major computer manufacturing firms had an advertising campaign, the thrust of which was the expression: *Not just data, but Reality!* This slogan was an inspired insight into the problems faced daily by those must make decisions. It captured the essence of the frustration that managers know when they are deluged with data rather than being given a distillation of the relevant facts they need.

Because data can be processed so easily, they are. The computer hardware and its supporting peripherals available to managers have capabilities that far exceed those of its users. But we are beginning to learn. This book has as its objective how to organize, control, and manage data effectively so that relevant facts are made easily visible when needed.

1.1 THE MANAGEMENT INFORMATION CRISIS

In 1978, 72 billion documents were processed as information. The number will escalate to 142 billion in 1988 and double again in another 10 years.

In spite of (or because of) billions spent for equipment to collect, process, and store data, an ever-increasing proportion of our total resources are employed to produce this information. The expense of administration and management has been one of the most rapidly escalating factors in cost of product. General and administrative expenses and general factory overhead expense have been growing disproportionately to direct labor cost growth.

Some of the disparity is undoubtedly due to the automation of manufacturing processes. But much of expense increase is due directly to the increased cost of processing and handling data. For example, in 1971, a major U.S. aerospace manufacturing firm found that more than 50% of its capital equipment worth was investment in computing hardware and supporting peripheral equipment (they included long-term leased equipment as owned). This shocker was revealed in crisis when commercial airplane sales were down due to a depressed market and a vital government-funded program contract was summarily canceled.

As useful and versatile as computers are, it still takes people to make airplanes (at least for a little while longer). The amortization of the investment in computers and lease costs, and the people to plan, manage, program, and operate them, are nonproductive expense. They are costs of administration, unless you sell the services of the computer to others for profit. The firm in question accelerated efforts to do just that. It was not planned that way, but that is the way they made it turn out.

1.2 PRODUCTIVITY AND DATA MANAGEMENT

With all documents generated to inform management, based upon billions of datum, decision quality should be at a very high level. The most effective measure of management and decision quality is productivity improvement. Studies of what has happened in the past 10 years reveal that U.S. managers are not doing too well.

Figure 1.1 was developed from figures published by the American Productivity Council. It compares U.S. annual productivity improvement from 1950 through 1974 with that of our major trading partners. The 2.9% annual productivity improvement rate of the United States is the lowest when compared with the rates of the United Kingdom, Canada, France, West Germany, and Japan.

Most economists and managers were not alarmed when they were shown these figures. After all, it was pointed out that (1) the U.S. productivity base was extraordinarily high to begin with, and (2) four of the nations represented were rebuilding war-devastated production plants as well as shattered economies.

Figure 1.2, however, tells a somewhat different story. It puts a new perspective on things. It tells it as it is. This chart was developed from figures published by the U.S. Department of Labor. It covers the period 1970 through 1975 and overlaps the first chart by 4 years. As in Figure

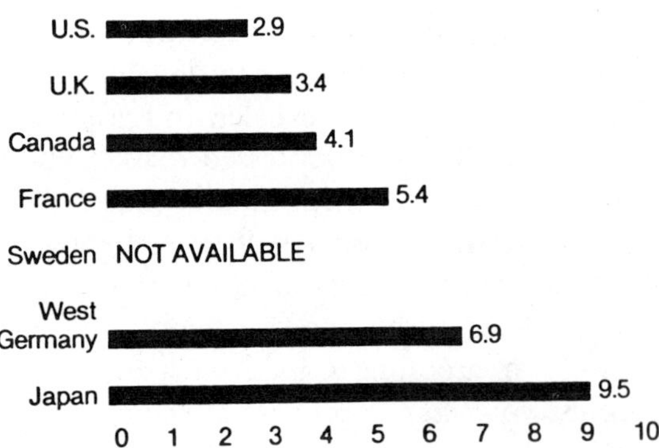

Figure 1.1 Average annual rate of increase (%) in manufacturing output per employee, 1950–1974.

1.1, U.S. annual productivity improvement is shown to be the lowest of its major trading partners. Although the productivity improvement rates of all the nations have slowed, there is no comfort to be found in that fact, and no excuses can be offered that are acceptable.

Our annual rate of productivity improvement averaged 1.8% per year during the 6-year period. The rate has dropped by 37.4% from the 2.9% average sustained over the 25-year period, 1950–1974. The Office of Wage and Price Stabilization forecasts that for 1978, the figure will be down to only 0.1%, or less.

What this all means is that if U.S. productivity improvement is taken as a base of 1, our major trading partners exceed our improvement rate as follows: United Kingdom, 168%; Canada, 168%; France, 189%; Sweden, 240%; West Germany, 300%; and Japan, 340%.

Although we are still the world's most productive nation, producing 28% of the world's goods and services with 5% of the world's population, our size and wealth of resources cannot sustain us in competition in a worldwide marketplace when productivity falls. We will be priced out of competition. Several nations will soon outstrip our productivity.

A good example of this can be shown by comparing U.S. productivity with that of France (see Figure 1.3). The productivity of France was approximately 85% of the United States last year. If France and the United States maintain their 4.3% and 1.8% annual productivity im-

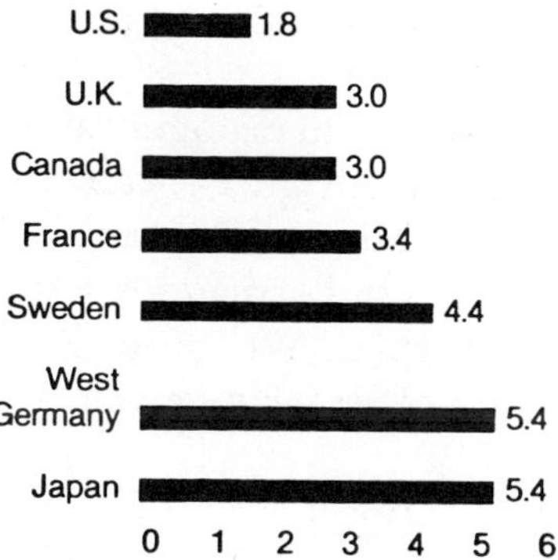

Figure 1.2 Average annual rate of increase (%) in manufacturing output per employee, 1970–1975.

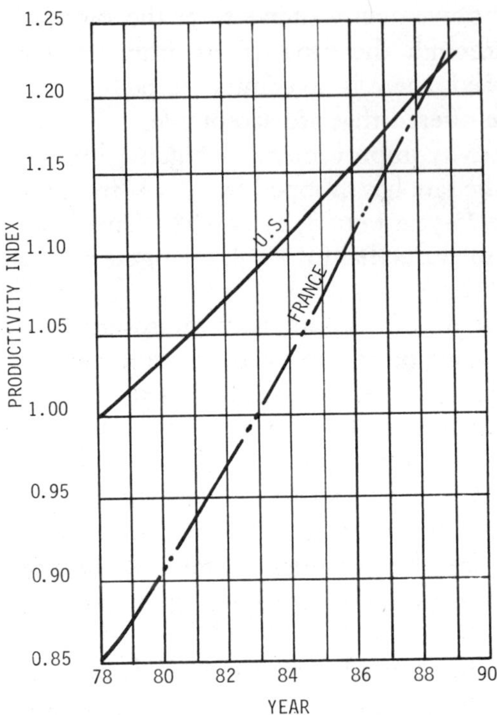

Figure 1.3 Productivity improvement comparison, United States and France (projected).

provement rates, France will equal U.S. productivity in 1988 and exceed it in the ensuing year.

Another measure of management is its capability to maintain labor cost stability per unit of product. The 20-year graph of Figure 1.4 reveals what has happened to labor costs per unit of output of manufactured product. During the period 1957–1967, labor cost fluctuations were modest. When smoothed, they are virtually constant. We can deduce from this that wage and salary cost increases were offset by improvement in productivity and/or cost reduction.

From the long-term graph of Figure 1.1, you will recall that that the *average* productivity gain for the United States was 2.9% annually. It is important to remember that this figure covered a span of 25 years, and that during the early years of that span, productivity improvement

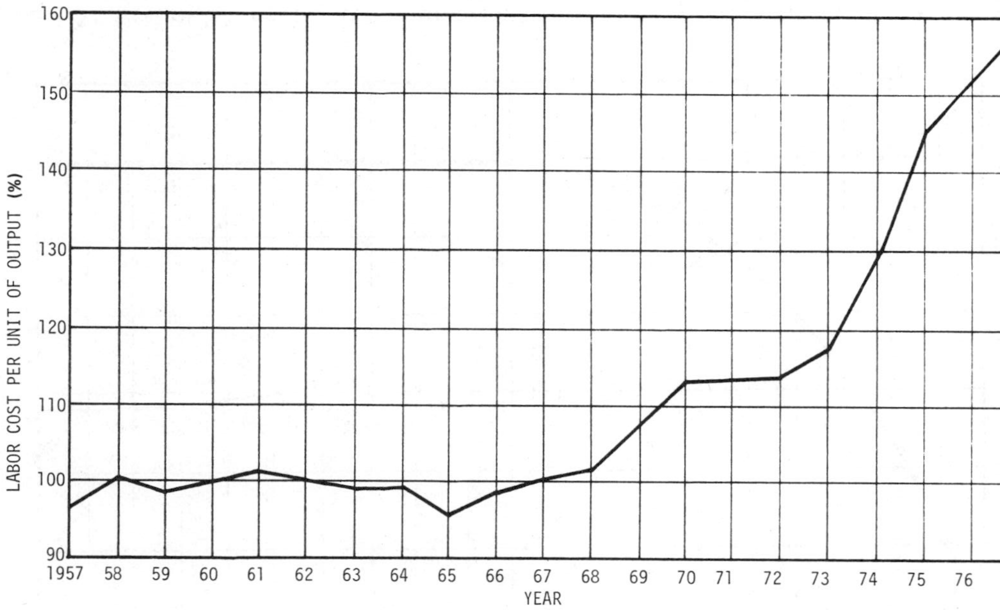

Figure 1.4 Cost of labor per unit of output, 1957–1977.

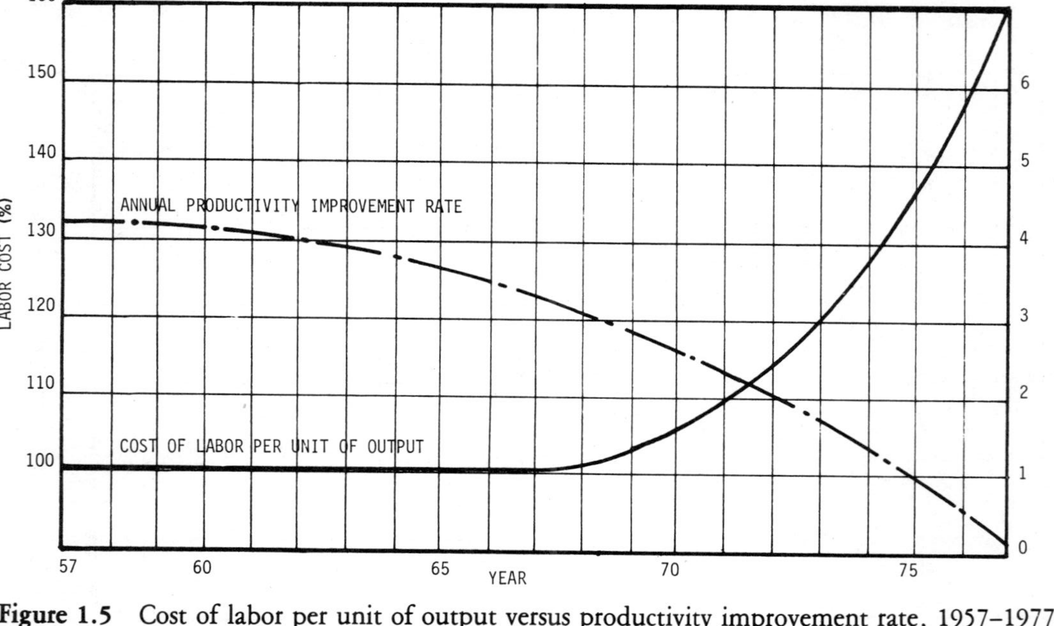

Figure 1.5 Cost of labor per unit of output versus productivity improvement rate, 1957–1977.

ranged between 4 and 5% each year. With a little smoothing, a composite of the productivity improvement charts and the labor cost graph can be shown (see Figure 1.5).

It does not require much genius to see that unless something is done to reverse the trend, we can expect the following:

The rate of inflation will continue to spiral.
The value of the dollar abroad will drop even lower.
Our export markets will shrink.
Imports will rise.
Our balance-of-payment deficits will grow even greater and the cycle will repeat.

It would be capricious to say that data management (or lack thereof) is responsible for these problems. No single factor can be said to be the principal cause.

1.3 THE PRODUCTIVITY/COST CRISIS: CAUSES AND EFFECTS

There are many and diverse causes for the drop-off in productivity. Some follow:

Cost-of living escalator clauses in labor agreements
The Vietnam war
Counter-productive taxation
Government inflationary spending policies
Nonproductive costs to comply with government regulations and agencies
Management reluctance to adopt new motivating incentives for American workers
Waste of resources on redundant work

1.3.1 Cost-of-Living Escalator Clauses

Some may recall the historic labor agreements reached by General Motors Corporation and its several unions in 1948 and 1950. They were historic for three reasons:

1. Labor and management acknowledged the desirability and mutual benefit to both parties of technological improvement to improve productivity and pledged cooperation in writing.
2. Management agreed to share the benefits of reduced costs by in-

cluding an annual technological improvement adjustment for the life of the contract.

3. The 1948 agreement included a cost-of-living adjustment factor, which could raise or lower wages regardless of the competitive and/or profit position of the company.

There was some good news and some bad news in these agreements. The good news was the mutual recognition of the need to reduce costs through technological improvement. That clause is frequently forgotten, with the result that there have been several work actions over work rules and productivity improvement.

The bad news in the earlier agreement is with us still. Rarely are wage agreements reached today that in any way reflect the productivity-offsetting aspect. Supply and demand and/or ability to pay fade further and further from the facts of the marketplace. Ask Chrysler!

It is a fool's game to grant cost-of-living wage increases and then pass the cost increase along. The results have been nothing short of catastrophic in the decade 1966–1975. Our unit cost of labor in manufactured goods has soared by nearly 60%. At the same time, our annual productivity improvement figure has been steadily diminishing. Shame on our management — shame on our labor unions.

1.3.2 Vietnam War

My only commentary on the subject of the Vietnam war is that we discovered too late that our guns-and-butter policy could only spawn a financial crisis of horrendous proportions. Its effect is still being felt and will continue for years.

1.3.3 Counterproductive Taxation

Of the six leading industrial nations in the world — United States, Japan, West Germany, France, Belgium, and the Netherlands — the United States is the one to have a capital gains tax. This nation has the lowest retention of capital by individuals of any of our major trading partners. Even the United Kingdom, with its repressive system of taxation, exceeds this nation in retained capital by individuals.

This has spawned an effect that impinges directly upon productivity. Managers of U.S. firms often do not have the capital necessary to modernize their production facilities. Estimates of this capital shortfall range from $200 billion to $300 billion annually.

Some of the investment capital shortfall is disguised because of investment from abroad. From individuals holding Eurodollars and firms suffering from revaluation has come the capital to invest in new plants.

Was it just 11 years ago that the Frenchman Jean-Jacques Servan-Schreiber wrote *The American Challenge,* warning the world, and Europe especially, of the economic aggression and capture of the world's productive capacity by U.S. multinational corporations?

1.3.4 Inflationary Spending Policies by Government

Governmental spending policies are a major contributing factor to the slowdown in productivity. The deliberate government-generated inflation has created a curious attitude that accepts this as a fact of life. J. Peter Grace, Chairman of W R. Grace & Company, in a 49-page booklet entitled, "The Disincentivization of America," claims that it is this politically expedient system of government overspending that has all but destroyed the incentive to invest in the American economy. He points out that investors who bought stocks and bonds in 1968 and paid taxes on the gain were left at the end of 1977 with 33 to 76 cents of the purchasing power of their original dollar. One dollar invested at the beginning of 1968 in the Value Line Average (a broad measure of stock prices) would have left the investor just 33.4 cents in buying power.

Mark the date June 2, 1978. On that day in 1978, you and I started earning money for ourselves. Until that date (January 1 to June 1, 1978) we were working for the government. In 1979, economists estimated that 35% of wages and salaries went for taxes. In 1980, the figure reached 41 + %, and it will reach 60% by 1990, unless checked.

The effect upon our economy of deliberate inflationary spending by government is well known and completely predictable. For more than 20 years, computerized econometric models of our economy have been used, with great accuracy, to track the cause and effect of policies in both the private and public sectors. Business school graduate students, working with simulated crisis situations, try to counter them by their decisions, and can track the result. Thus, it can only be concluded that what is being done is being done deliberately. If not, a crash course in basic economics is highly recommended, and *Das Kapital* and *The Wealth of Nations* should be mandatory reading in Washington.

1.3.5 Nonproductive Expense for Compliance with Governmental Regulations

As relevant as "disincentivization" reversal is the need to clear out overlapping, conflicting, and capricious regulations. Inbred, distorted viewpoints of bureaucrats far removed from the real world of competitive business have created a burdensome expense that bodes ill for everyone.

Although there are no national statistics available as to the effect of such regulations upon productivity figures, several firms have stated that from 18 to 28% of general and administrative expense is devoted to:

Filling out city, state, and federal tax forms
Collecting taxes for these government agencies
Answering questionnaires for various agencies
Collecting FICA
Complying with ordinances and regulations to limit noise, pollutants, hazards (OSHA), employment regulations (EEOC), pension regulation (ERISA), ad infinitum, ad nauseum. Some of this may be necessary; most of it is not.

1.3.6 Management Reluctance to Adopt New Motivational Incentives

Management has been slow to recognize and adopt a new order of thinking about precisely what motivates people to be productive. These new paradigms were researched and identified by Abraham Maslow of Brandeis University more than a decade ago. Frederick Herzberg expanded upon the work of Maslow. Later, Robert N. Ford adapted these theories to practical, working procedures.

Management and industrial engineers alike must conclude that some of Taylor's postulations have become obsolete. The "one motion/one job" objective of ultimate simplification of work is illogical and unacceptable to today's worker.

Prior to Maslow's work, Peter Drucker* identified the problem. Drucker's work was not couched in psychological terms; he directly criticized managers and management engineers. In his 1954 book, *The Practice of Management,* Drucker pointed out the virtual bankruptcy of personnel management. He highlighted the failure of scientific management to help us with the problem of managing workers and work.

Taylor could not possibly have foreseen the advances in work measurement techniques that would develop, nor could he have envisioned the qualifications of postwar American workers. Drucker said

*P. Drucker, *The Practice of Management,* Harper, New York, 1954.

that with the possible exception of Lillian Gilbreth and Harry Hopf, there had been no enlightenment of note since Taylor, Gilbreth (Frank, Gantt, and Hathaway. Remember that Drucker's book was published prior to the publication by Maslow of his findings.

Predetermined time systems gave management and industrial engineers a means to analyze the basic motions that comprise any method of doing work. But the fragmentation of work has become so fine that work no longer offers a challenge. Maslow proved that fear no longer motivated, as it did in Taylor's day, and that money never really did motivate—at least not in the modern context.

Harry Hopf saw the value of work integration, as he called it. So did Maslow, Herzberg, and Ford. But it was the Scandinavians who grasped the full significance of what was meant. In Sweden alone, more than 500 companies* have redesigned jobs to put challenge back into work—challenge to the whole person, not just hands, back, and feet. They saw that the Taylor principle of the division of labor (i.e., separation of the planning of work from the doing of work) is akin to separating the head from the body. To nourish the body requires the head.

More than 25 years ago, IBM had proven that productivity greatly increased when the responsibility for planning the work was given to the people doing the work. What does this entail?

Programming the equipment
Setting it up
Maintaining it
Scheduling the work
Doing it
Checking its quality

It is evident that the worker of today has far more capabilities than his forebears had. To continue to waste this resource by trying to make robots out of flesh and blood—thinking human beings—is not only counterproductive, but it is at the root of much of the discontent of our working force.

Daniel Yankelovich, in an excellent article entitled, "The New Psychological Contracts at Work," which appeared in the May 1978 issue of *Psychology Today,* claimed that we must rethink our value systems as they relate to motivating people at work. Yankelovich postulates that we

*Swedish Employer's Federation Report, 1975.

must reverse the trend of depersonalization if work is to be meaningful. A paid job of *challenge,* involving decision-making processes, is more important than money as a motivator.

Although managers frequently extol the value of "our people," attention is paid to everything but people. It is a burden that most managers find disgruntling, but disgruntling as it may be, this is one challenge that can be turned into a great opportunity. With the tools available, and the knowledge of how to rebuild the work so badly fragmented, management can make this not only a more productive economy but also a better society.

Data classification and coding are tools that enlarge the range of human vision and permit decisions to be made at a lower level.

1.4 SOME SUGGESTIONS ON PRODUCTIVITY IMPROVEMENT

Productivity must be improved—through

1. Classification and coding of our resources, to control unnecessary variety proliferation
2. Group technology, to reduce resource input to yield greater efficiency
3. Job redesign, to motivate the new breed of American worker

The unnecessary proliferation of data, parts, materials, tools, and machinery is a waste that none can afford. Once the least common base of data has been achieved, we must ensure, through workable standards and controls, that redundancy is prevented from occurring in the future. No one will argue that this goal is undesirable. Yet this country is just awakening to the fact that billions are squandered annually by not taking full advantage of this important, productivity-improving source of profit.

Standarization has been a workable tool of European managers for 60 years. Small lots were their way of life because their markets were smaller than ours.

Present-day classification, coding, and group technology are European imports, but classification and coding were used by Taylor at the turn of the century. Taylor's classifications were used to organize, make visible, simplify, and standardize the component elements of businesses as long ago as 1902. In 1904, for example, Morris L. Cooke, a pioneer in the then-new management, said: "Only as we learn to classify and to code, to simplify and to standardize, will any real science of management emerge." In this country, we took this to mean concentration on only one of the resources of business—the human resource—because in Taylor's time that was the greatest single cost.

The Europeans, however, directed their attention to other resources. In the USSR, Sokolovski developed his postulation regarding the standardization of the technological process, which was that things that look similar should be made in the same way. S. P. Mitrofanov, also in the USSR, built upon that postulation to develop what we know today as *group technology.*

Classification, coding, and group technology are all approaches to solving the problem of redundancy in management time, money, material, and machinery. The ***human resource was not the principal*** *target for control.*

Our firm, E. G. Brisch and Partners Ltd. (now Brisch, Birn and Partners), got its impetus in the control of the component elements of business and industry during the latter stages of World War II. The firm was founded in 1948 to use classification and coding to help management control redundant effort and curb unnecessary proliferation of *all* resources. By bringing like things together and making them visible—then, and only then, can variety be managed.

In May 1978, at Coventry, England, the thirtieth Anniversary Conference on industrial classification and coding was held. It was most gratifying to hear the results achieved during these 30 years by several of the firms. One speaker, representing the world's largest producer of diesel engines, told of controlling raw material variety over a 17-year span (until metrication) so effectively that they had added fewer than 1% of new items during that period.

Another speaker reported that in a 16-year period his firm had controlled the influx of new commercial parts, raw materials, and new designed parts so effectively that tooling budgets had been dramatically reduced in spite of a fivefold real increase in sales. The firm credited a 14% profit on sales and a 26% return on invested capital to the effective use of classification and coding and group technology.

1.5 DATA ORGANIZATION

Managing data begins with the establishment of an order. This is a hierarchical arrangement of the parameters each datum has in common with all others so that like data are drawn together. It is a logic tree, a branching method in which each decision is binary.

The data we refer to are factual data. They have undergone discrimination and test after capture. The levels of data to be considered and the files in which they are to be stored are as follows:

Raw data: As captured.

Source data: Data remaining after redundant data are extracted and erroneous or ambiguous data are properly identified.

Classified input data: Source data that have been organized into a hierarchical order in which like data are found together in families.

Coded input data: Classified source data which have been classified and codified to reflect their family identify and their given unique identity within the family group.

Classified and coded file: A collection of classified and coded data items stored behind record file addresses which have been organized to produce special reports.

Classified and coded master file: A collection of semipermanent data items in record order for repetitive processing of reports.

The essential step for report reliability is data quality control, which results from the classification and coding process. Without this essential step, the entire controls system becomes suspect. One of the earlier and much abused acronym expressions coined by data processors for data integrity is GIGO, garbage in, garbage out. Another comment is perhaps more relevant: "I am constantly annoyed at the numbers of people who code erroneous data and believe that when it is decoded, it will somehow have been purified in the process" (from a paper delivered by George Melville Dewey in 1878).

Each file must be tagged with a key to identify it for processing. Special records label the file at its beginning and end. Our direct concern is with the address of the records (on film, tape, disk, or hard copy) within a file. We will deal with the advantage of classified and coded records within files regardless of the storage method utilized.

Where a file is accessed for processing, it is essential to know the file address as well as the file name. Without this information, the required data are obscured and cannot be retrieved. For example, if the computer is to print the process routing sheets for all parts and assemblies on the schedule, without the file name and record addresses, the data cannot be located and retrieved.

Circumvention of this problem requires a subroutine to find the proper codes.The subroutine may be programmed in a central processing unit, in a satellite computer that interacts with the central processing unit, or it may involve a manually operated procedure. Regardless of the mode of operating, the classified and coded data file is what makes it possible to find the relevant file addresses.

1.6 DATA RETRIEVAL PROBLEMS

Many firms with computer installations of some maturity have not solved the problem of data retrieval when the file address is unknown. This is especially true in the product design and manufacturing engineering functions. Without this capability, the computer cannot function as effectively. Earlier in the chapter, reference was made to redundant data and the decision-making problems they can cause. If data cannot be found, they cannot be used.

For example, one computer manufacturer had 11 different part numbers for the same 1% metal film resistor. How can a sophisticated product manufacturer with high-technology engineers find themselves with such a problem? The parts identification system was a sequential number of seven digits assigned in random order. The filmed drawing files were maintained in ascending numerical order. To determine whether there was an existing resistor that satisfied the designer's requirements meant at least a half day's search of the files. It was far simpler and faster to have a new number assigned, and thereby create another record.

Memory is unacceptable as a means to retrieve, even though there are memory marvels who seemingly never have to search for data. Yet there are those who propose standardized terminology teamed with a key word in or out of context systems as a solution. Very few installations have been found to be workable without the data first having been identified and classified.

Since memory cannot be depended upon, how can tens of thousands of file records be accessed to find either the relevant file and/or the data items needed for a decision? F. W. Taylor was a strong advocate of classification and coding as a means of managing the myriad facets of a business. At a meeting of the Taylor Historical Society in 1920, H. G. Hathaway* quoted Taylor as follows:

*H. G. Hathaway, colleague of Taylor at Tabor Manufacturing in 1904–1905.

> A classification is essential to an orderly arrangement of the facts relating to a business and to the orderly conduct of its activities. During the period of development and installation of a system of scientific management, it is especially helpful to a proper visualization and understanding of the business and its problems, as well as to the conduct of the work.

Taylor approached classification as a scientist in management. He used the tools available to scientists and scholars at the time. The Dewey Decimal System,* a library system, was his choice to classify and codify the data items of his day that related to business and industry.

Dewey, in a pamphlet published in 1876, divided human knowledge into nine classes—.1 through .9. He reserved .0 for newspapers, encyclopedias, and other material too general in scope to be assigned a specific class. He assigned the number .6 to all *applied sciences.* Business, one of the applied sciences, was assigned .65 and the management of business, .658. The management of the manufacturing function is .6585. The time and motion study subfunction is .658542, while an individual time study would be coded .6585421.

Taylor altered the system by dropping the decimal point, the Dewey trademark, and used a colon to show the relationship of one code to another. The woodworking industry, for example, is 674. To identify time studies on a woodworking machine operation such as a double-end tennoner, the code would read 674:65854216.

The code is open-ended and varies in length. Taylor quickly recognized that in order to code a cutting tool, the number of digits would extend the code to such a length that it would be virtually impossible to transcribe it without error. He, therefore, developed generic subcoding systems for limited populations of things that could be used by several industries. He also departed from the Dewey system by using mnemonics. His code for drills, reamers, and taps was used by the National Twist Drill Manufacturers' Association from World War I through World War II (e.g., ¼ HSSS identified ¼-in.-diameter high-speed straight shank). The proliferation of items, materials, and processes generated by the war effort expanded the code enormously.

Another pioneer in industrial management, Frank B. Gilbreth, applied classification and coding to the basic divisions of human endeavor. His

*Ibid.

therbligs (Gilbreth, spelled backwards) made it possible to analyze, classify, and code all human motions that could be used to perform any work.

Gilbreth postulated that the motions made by a human being are finite in number (originally, there were 19 motion therbligs and 1 hold therblig) and are constant. Subjectively developed by Gilbreth (assisted by his wife, Lillian M. Gilbreth), later studies by a special committee of the American Society of Mechanical Engineers (ASME) simplified the number of therbligs, combining several that always occurred together.

A. B. Segur combined and refined the work of Gilbreth and Taylor when he developed a classification and code for motions by the body members with predetermined times for each motion. Segur postulated that the time necessary to perform any motion by an expert was constant. His Motion-Time Analysis procedure, a product of the 1920s, is recognized as the forerunner of other predetermined motion-time systems (also called predetermined time standards), such as Methods-Time Measurement (Maynard, Stegemerten, and Schwab) and Work Factor (Quick, Shea, and Kohler), both of which began developing about 1940 and were publicly announced in the late forties. Subsequent generations of classification and coding for predetermined time systems simplified the work of these six. One such system, Master Standard Data (Crossan and Nance), combined the most frequently occurring motion sequences into larger blocks of time.

Figure 1.6 graphically portrays the evolution of classification and coding systems for the purpose of *measuring, simplifying,* and *standardizing* human endeavor. It spans the period from Taylor through 1964. Since the printing of this illustration, even more systems for the same purpose have been developed and made available to management engineers.

Just about the time that Work Factor and Methods-Time Management were being introduced, modern industrial classification and coding for *all* management data was born. In October 1948 in London, two ex-patriot Poles, E. G. Brisch and Joseph Gombinski, founded the first firm in the world dedicated solely to the control and management of data. Their concept combined the classification methods of Linnaeus, to establish logical data order, with cryptology, to establish systematical storage, retrieval, and control. This concept, in practice, makes it possible for any manager or administrator to access and retrieve any data or facts relevant to a decision—tactical or strategic—instantly and easily.

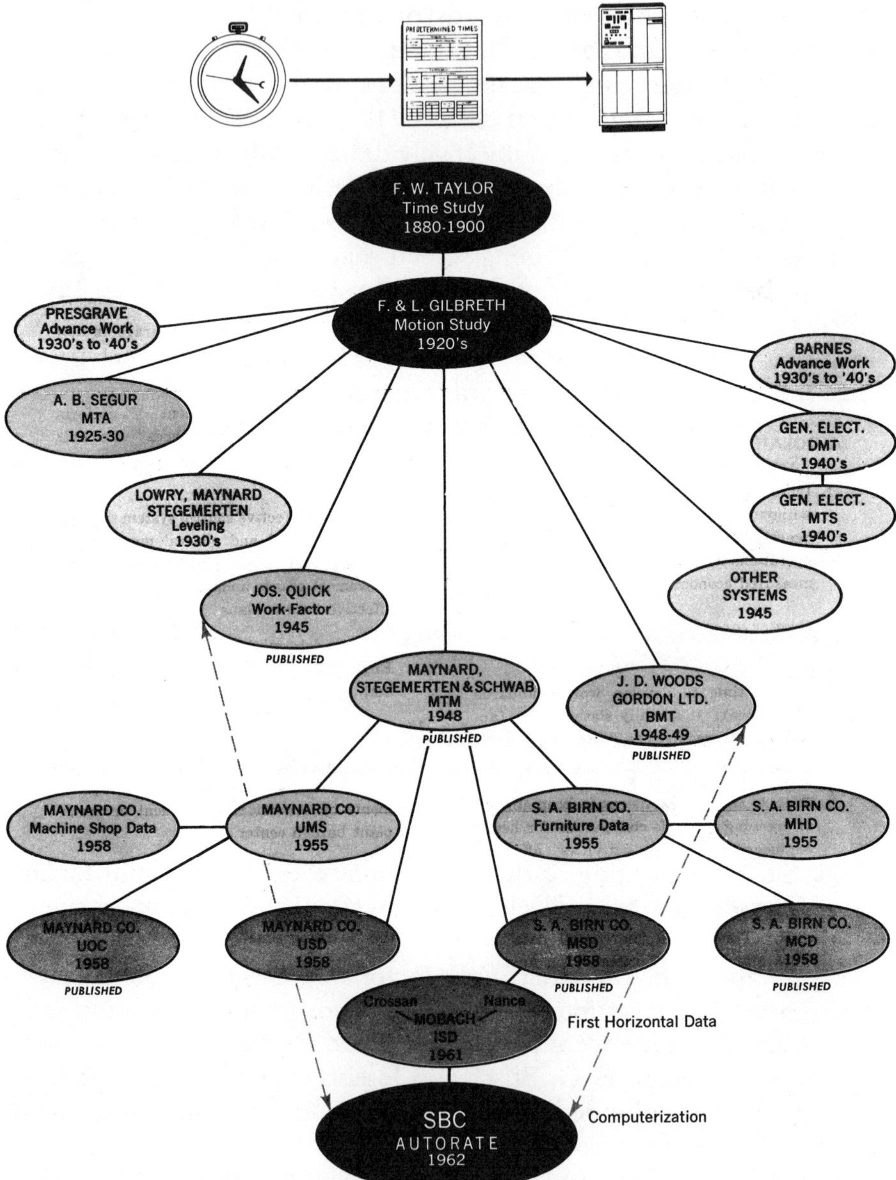

Figure 1.6 Major developments in the history of work measurement leading to AUTORATE. For the sake of brevity, many other contributions could not be included. (Reprinted for Serge A. Birn Co. through courtesy of The Service Bureau Corporation, New York.)

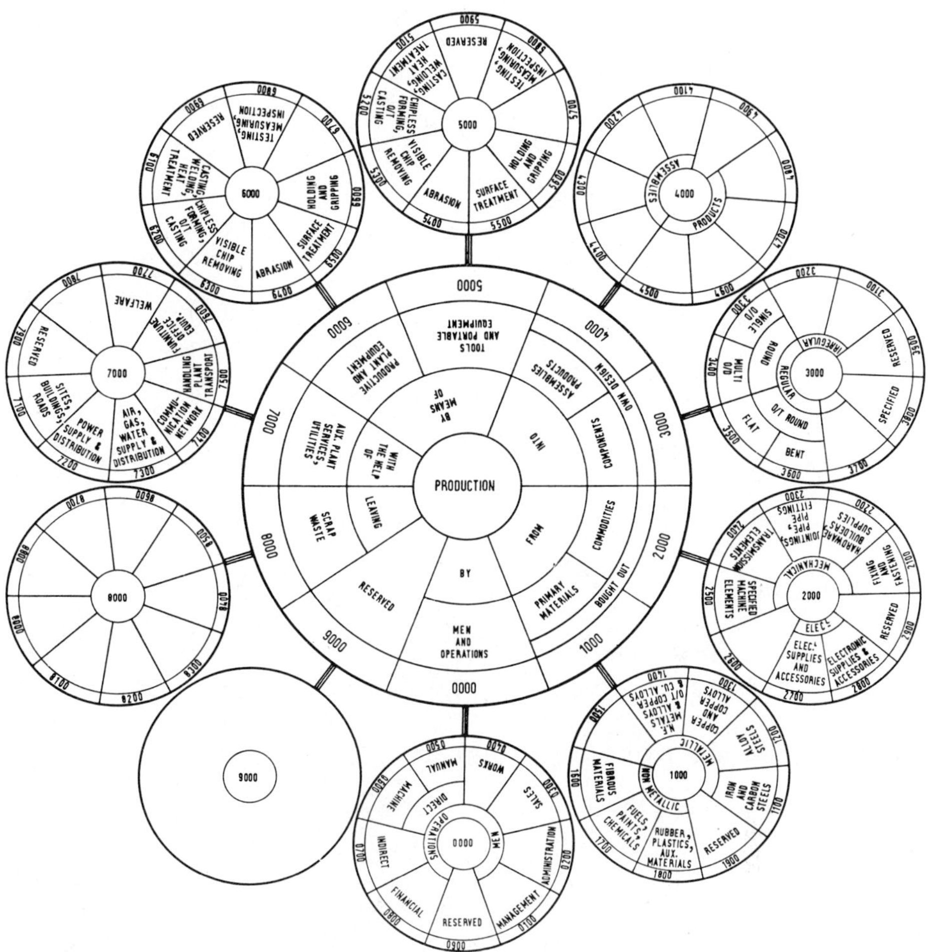

Figure 1.7 Graphical concept of the component elements of production.

The categories of data that might typically be necessary in the production of a product are represented graphically by Figure 1.7. This cosmic view of the classes of data to be considered for a classified and coded data base makes possible work on any part or the whole, without prejudice. Figure 1.7 is a hypothetical example of an engineered product manufacturing firm. Neither the classification nor the codes illustrate all the universes or all applications of industrial classification and coding. This Brisch Wheel, as it is called, is not meant to be a universal model for all of industry, business, and government.

2

Not Knowing the Cost of Out-of-Control Data Can Hurt You

2.0 SOME FACTS OF LIFE

Contrary to the old adage, "What you don't know won't hurt you," what you don't know about the cost of redundant data proliferation *is* hurting you, whether you know it or not. It hurts your managerial productivity and dissipates profit.

The productivity of managers is measured by the number and soundness of the decisions they make. A data base not in control, not defined, not identified, not classified, and not coded for easy retrieval impinges unfavorably on this quintessential contribution.

In *The Practice of Management,** Peter Drucker said:

> Get the facts is the first commandment in most texts on decision making. But this cannot be done until the problem has first been defined and classified. Until then, no one can know the facts; one can only know data. Definition and classification determine which data are relevant; that is the facts. They enable the manager to say what of the information is valid and what is misleading.

If you were to canvass your managers, few would associate the quality of the decisions they make with the capability to find relevant

*P. Drucker, *The Practice of Management,* Harper, New York, 1954, p. 358.

facts from all the available data. Fewer still would be able to identify their specific data problems, and practicaly no one could put a price tag on the sum of the problem, overall. So suggesting the idea of controls for a problem few appreciate is not likely to be greeted with much enthusiasm. The exceptions will be the few who have suffered from the problem long enough to know how it can hurt, if not how much it is actually costing.

2.1 HOW TO GET MANAGEMENT'S ATTENTION

Unless you happen to be the chief executive officer or chief operating officer of your firm, it is not always easy to gain top management's attention to a problem. Just as in training mules, if you are to get results, you first have to get their attention. Rather than applying a 2 × 4 smartly across the nose, we recommend that you present hard facts on what it costs your firm each year to support the habit. When the costs of not having data under control are known and properly documented, you will capture management's attention.

There is nothing quite like finding a new and yet untapped source for cost improvement to get a favorable decision from the board of directors to invest capital. That is, of course, if the cost to realize the benefits is in a favorable ratio to the savings.

2.2 MEETING PAYBACK GUIDELINES

Capital investments, whether they be for new equipment or new administrative management systems, should be cost-justified. Investment guidelines are almost always established by the board of directors and must be satisfied if a request for program funding is to receive favorable consideration.

These guidelines may be expressed as an after-tax contribution to profit, say 25–30%. They may be expressed in years to reach the break-even point, say 2 years. They may be expressed as a before-tax return on investment percentage, say 50% per annum. Some firms require a projection payback rate for each year of the expected life of the program. Some require break-even charts, and so on.

The order of business, therefore, is to determine:

1. The nature of the data control problem

2. The cost of having the data out of control—or, alternatively, what data control will yeld in savings
3. The cost to develop and implement data control
4. The return on investment, expressed in the form and terminology necessary for board/management acceptance of the program funding request

2.3 DEFINING THE PROBLEM: WHO?

When an organization is polled as to the one most critical company-wide data management need, each functionary will probably give a different answer. Invariably, if an opinion is expressed, it will often reflect the parochial problems of the respective respondent. Therefore, any examination of the data control problem(s) and priorities for solutions must be made by objective people with a view of the organization as a whole. This may be done internally by qualified staff people, provided they are free of daily operating pressures and/or independent of functional staff responsibilities for the duration of the study.

The study may also be conducted by a qualified external group or a combined team of internal and external people. In all cases, the key is objectivity. The problem must be viewed dispassionately and free from bias—that is, as free of emotion and personal prejudice as is possible.

Once the team of surveyors has been selected to make the study, it is necessary to provide a charter. This means establishing the objectives for the study and communicating them to the organizational functionaries who will be involved. How this is done can best be illustrated by relating an actual case history.

Ten years ago, a small firm of about 350 people in the fluid-valve-making business found that it had a problem related to accepting growth. This firm made a variety of valve types for defense and industrial application, especially for nuclear power plants and nuclear-powered submarines. A labor-intensive product, most of the product value is added in-house. As the variety of new products began to proliferate, coupled with the retirement of several key people, problems with data management began to surface. A computer was decided upon as the way to solve the problem. But this decision raised another problem. The existing part numbering system was inadequate.

In-product raw materials were not identified by part number, nor were many of the commercial commodity items, such as fasteners, bearings, seals, and the like. Both categories were described by name on the bills of materials (BOM). Drawings of proprietary designed parts did not show the materials from which they were made. Each part that came in contact with the various fluids being regulated were made from a variety of materials, according to application or intended service.

Bills of materials were prepared by hand. They identified each designed part by its drawing number and the name of the material from which it was last made. This meant that on the next order, if the same valve model and its parts were to be for a different service, they very likely would be made with a different raw material for contacting the fluid. The bill of materials master would be physically changed by erasure and the new material added. No record remained of prior materials used. This created many service problems, as there were more than 10,000 different raw materials carried in stock, plus several thousand more purchased upon demand.

To develop a computer-based bill-of-materials system, it had been suggested that a six-digit nonsignificant identifier tag number be assigned to each part, raw material, assembly, and product. The computer manufacturer had recommended this as the fastest and least expensive way to get *one number for one thing* into the file. The data, however, were so poorly described and so inaccessible that grave concerns were expressed by the chief engineer that this method could work. He had concluded that most of the problems would simply be transferred into the computer file, become invisible, and never be resolved.

The controller was responsible for computer systems and data processing. He reluctantly agreed to study the problem more extensively before making a final decision regarding the part numbering. His reluctance was understandable; a computer was standing partially utilized, awaiting the outcome.

The controller appointed an internal study task force of four (one person from engineering, one cost accountant, one systems analyst, and one production controller) and the controller served as chairman. Several fruitless months went by, characterized by endless interruptions to the study. The team had tried to do justice to their regular jobs while conducting the study. Finally, the controller disbanded the internal team and commenced to interview outside firms which might resolve the problem more quickly. One firm was selected from several interviewed and the study was begun. The systems analyst from the prior team acted as the liaison person.

A program plan for the study, together with the data that were to be developed, was presented to the management of the valve firm in an introductory meeting. This meeting explained the study objectives. It also served to introduce the surveying team and define their charter, and to obtain top management's approval to be given access to all necessary data. The following is a list of the data needed to conduct the study:

I. The number of sizes, types, and kinds of:
   A. Primary (raw) materials.
   B. Purchased nonproprietary parts and assemblies.
   C. Proprietary piece part designs/drawings, created in each of the preceding three years, including those:
      1. Designed, but not released.
      2. Designed, released, and produced.
   D. Proprietary assembly designs/drawings, with the same qualifications as in (C).

II. The number of purchasing transactions executed in the three fiscal years immediately preceding.

III. The number of shop orders issued in the preceding fiscal year.

IV. The average value of inventories of:
   A. Raw materials.
   B. Nonproprietary parts and assemblies.
   C. Work in progress.

V. The annual cost to carry inventory, expressed as a percentage of value (exclusive of adjustments due to appreciation) and made up of the following factors:
   A. Interest charges on the investment (1% above prime rate will suffice).
   B. Warehousing and accommodation costs.
   C. Deterioration and obsolescence.
   D. Taxes, if applicable.
   E. Insurance.
   F. Initial receipt.
   G. Mysterious disappearance.

VI. The cost to execute a procurement from the request to buy through reconciliation of the paid check and invoice.

VII. The inventory turn rate (total procurement divided by average inventory).

VIII. Evaluation of present parts identification systems.

IX. Measurement of the effectiveness of engineering standards as a means to control variety systematically.

X. Documentation procedures.

XI. The cost to create new parts and assemblies designs, including the following factors of expense:

   A. Product design
      1. Design development
      2. Engineering
      3. Drafting
      4. Documentation
   B. Manufacturing engineering
      1. Processing
      2. Tooling
      3. Methods and work measurement.
   C. Production and inventory control.

   The costs of each of these is to be developed by relating the annual expense of salaries and fringes to the number of new parts and assemblies released annually to production.

XII. Productivity improvement of the preceding functions resulting from reduced information research.

A sampling was taken of every possible data source. This included data from order files, bills of materials, parts drawings, process routings, production and inventory control cards, traveling requisitions, purchase orders, cost accounting records, and even the bin cards in the storerooms.

Many closely similar and duplicated materials and designs began to surface. This is one of the most costly manifestations of not controlling data input to the files. Only 4000 of the 10,000 raw material items in stock were necessary to operate the business.

Eleven percent were duplicates due to quirks in identification. One example that illustrates this point concerned ¼-in.-square woven metal-cored asbestos packing. In the study of the inventory control cards, there were two records, one with maker's number, 6 AM, and one with the number 6 AM-CR, same maker. The surveyor examined the material in the stockroom bins. Experience with this maker's products and a study of catalogs made him suspicious of the distinction. He went to see the materials in the storeroom and found two bins, one for each material designation. The bins were 4 ft apart, with several bins between them. Each held boxes and materials that appeared to be the same. The boxes in one bin had a masking-tape band on which 6 AM-CR was written by hand in marker ink. The other bin held boxes printed with the maker's product designator, 6 AM. Except for the marked masking-tape band, they were identical.

The insurance-stock level was 60 lb in each bin at $40 per pound each ($4800 total). The surveyor had examined the maker's catalog and found only the 6 AM designation. However, there was a customer option note: 6 AM could be supplied in Cut Rings or Continuous-Roll material, so the CR identity problem was still not resolved.

The surveyor called the firm's local distributor. The distributor had no had no ready answer. He did say that some of the purchase orders from the firm being surveyed called for 6 AM, whereas others called for 6 AM-CR. He added that if the boxes did not include the designation 6 AM-CR when the purchase order called for it, the shipment was returned as rejected. So the distributor's people had taken to adding the tape and marking it 6 AM-CR. Then when the same order was redelivered, it was accepted. He was as puzzled as the surveyor about this until he recalled that, some 8 or 10 years previously, the maker had supplied 6 AM in either black iron core, as standard, or with Corrosion-Resistant core—6 AM-CR. But at least 8 years prior, the manufacturer had eliminated black iron core, switching to corrosion-resistant core wire as standard and thus dropping the CR suffix.

When this and numerous other examples were quantified and made known to management, the decision was not just to tag-number existing materials reputed to be different, but to identify, classify, codify, and simplify the variety of their in-product components and materials.

It was a justifiable investment. The survey study revealed a variety reduction potential of nearly 40% of the existing purchased materials and 12% of designed parts and assemblies. After determining the costs attendant with getting these data under control and manageable, it was found that the return on investment would be 275% the first year and 110% each year thereafter. In fact, these figures took a little longer to realize than planned but have since risen well beyond the original estimates. Other cases reported later in the book show comparable results.

2.4 COST-JUSTIFYING A DATA CONTROL PROGRAM

When there is more than one opportunity for capital investment, several programs must compete for the available capital. This means that a program for improving data management must be quantified to the fullest extent possible—not just the costs involved, but the return on the investment as well—if the program is to have a chance.

Determining the expected costs is simple, straightforward, and reasonably accurate (within a range of ± 10%). The projection of savings opportunities is not as straightforward. This is because some savings are cost avoidances. Some are quite difficult to quantify, yet are tangible sources for profit improvement. Some are just not quantifiable at all, but should be listed anyway.

To illustrate—a program to reorganize personnel data for an improved capability to develop managers (management development) for promotion-from-within replacement is virtually impossible to quantify as to profit improvement potential. It is possible to show a reduction in time necessary to sort through data manually to match position descriptions with employee qualifications. But this is, at best, a relatively minor savings, even with substantiating clerical measurement to prove the figures.

There are other benefits, but they are not credibly quantifiable. What, for example, is it worth to place the best candidate in a key position that has become vacant? If the prior manager was doing a first-rate job, it may mean no more than maintaining the status quo. To try to evaluate what effect a poor candidate might have upon the department he or she is to head would be suspect, pure conjecture, even though increased turnover and less efficient operations are themselves measurable.

Managing data more effectively in other activity areas of an organization can be measured more credibly. To illustrate, take data that relate directly to the product and its components from design through manufacture. To determine the savings and/or cost improvement is reasonably straightforward and can later be audited for adherence to the projection.

2.4.1 Purchased Materials

Let's use as an example in-product raw materials and commercial nonproprietary parts and assemblies that are bought either to stock or to shop order. Items bought to place in inventory incur not only the cost of procurement, but accumulate other costs as well, just by being held in the inventory. Investment in inventory must yield a return just as any other resource, over and above these other costs. Therefore, the cost to carry inventory must be known. The costs to carry inventory annually are both finite and quantifiable.

When the data (the records representing each individual item in the inventory) are identified, classified, simplified, coded, and cataloged, the variety represented is reduced. It has been the experience of many who have done so that, from 30 to 40% of the numbers of different kinds, sizes, and types of items existing prior to a data-control program will be eliminated over the course of several years. This will have a direct impact upon a number of cost factors, as well as upon the amount of operating capital invested in the inventory.

Figure 2.1 was empirically derived from experience during the preceding 30 years with firms that have solved specific data management problems of this type. As can be seen, a 36% reduction in inventory variety, as an example, will result in a 12% reduction in the investment in inventory. This reduction in variety (and, in turn, in investment) does not adversely affect service level. Quite the contrary. Service level invariably improves (fewer stockouts per total transactions) and, because there are fewer items to manage, there are fewer bought-to-shop-order items to track and expedite.

The variety versus investment-value relationship is not difficult to understand if the example given earlier is used to illustrate the point. As you will recall, there were two bins, one each of 6 AM (item A) and mismarked 6 AM-CR (item B). Each had 60 lb of safety-level stocks at $40 per pound. In this case, because they were duplicates, each application of both items was exactly interchangeable without debate. However, variety reductions are not limited to the elimination of duplicates. Duplicated items account for only about 4 to 8% of the total. Simplification of unnecessary, closely similar varieties will reduce the number of items by another 24–32% of the total. Disposition of duplicated items can be handled quickly and easily, as can some of the closely similar unnecessary variety items. Here it is necessary only to accept the premise, but more will be said on this subject in Chapter 5.

The point is that there was 120 lb of the same material held as safety stocks for two separate items to maintain a planned service level of X percent. To maintain that level of service for the combined use of items A and B, only 60% of the 120 lb (or 72 lb) is necessary. At $40 per pound, this reduces the inventory value by $1,920. So instead of an investment in safety stocks of $4800, now $2880 will suffice, without increasing the chance of a stockout. This frees $1920 (one time) for other purposes.

There is also an annual cost avoidance by not having to carry the superfluous $1920 of inventory each year. This "cost to carry inventory"

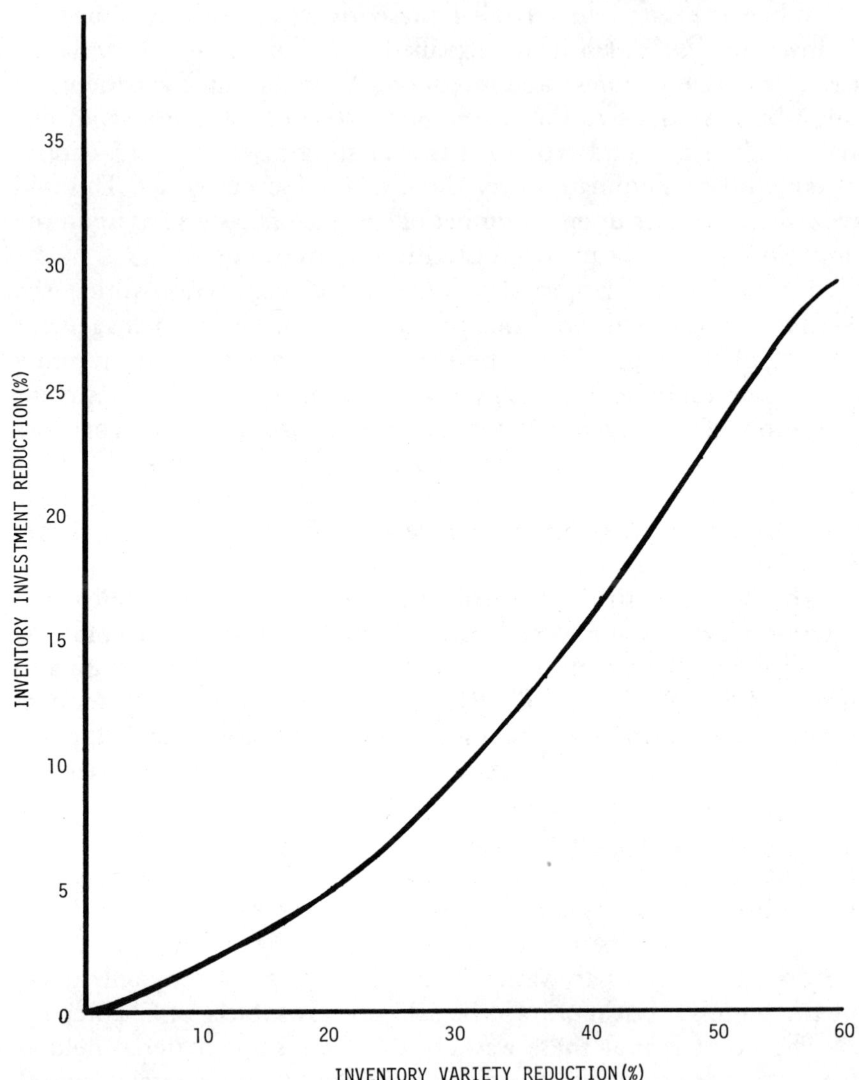

Figure 2.1 Inventory variety reduction versus inventory investment reduction.

figure is frequently debated. There is lack of agreement as to whether it is a variable cost, a fixed cost, or both fixed and variable. In any event, the constituent elements of the cost of carrying inventory, expressed as a percentage of value, can be stated. If economic order quantities are used in stock replenishment, the cost to carry inventory must be accepted as a valid cost, because the principal (offsetting) factors that must be balanced are the cost of a stockout versus the cost of carrying inventory. Therefore, the constituent elements of the cost to carry inventory are as follows:

Charges for the capital invested. There are two schools of thought on this. One school insists that money invested in inventory, if put to other uses, would yield a much greater return than the cost of borrowing. The proponents who hold to this school point out that any capital invested must satisfy an after-tax contribution to profit of, say, 20–30% per annum.

The other school takes the position that money is usually borrowable at close to the prime rate (according to the credit rating of the firm, of course), because the loan is secured by the inventory, and that not all other capital investments can be so financed.

The second position is more conservative and is recommended. So if your firm has an excellent credit rating, use 1½% over the prime rate as a reasonable charge for the use of the money value of the inventory. In recent years, the prime-rate figure has fluctuated widely ranging from as low as 6% to as high as 20%. If, for example, 7¼% is the prime rate, use 8¾% as the annual interest charge on the use of the capital.

Taxes on the inventory as a percentage of inventory value.

Insurance on the inventory as a percentage of inventory value.

Warehousing and accommodation charges (i.e., floor space, utilities, and equipment), expressed as a percentage of inventory value.

Allowances for deterioration, obsolescence, and mysterious disappearance (taken from the year-end inventory adjustment figure worksheets) divided by the inventory value and expressed as a percentage.

The sum of these calculations will usually fall within a range of from 17 to 25% of the total value of the inventory. What this means is that you are paying from \$170,000 to \$250,000 each year for each \$1,000,000 in inventory. If this calculation is more than you wish to undertake, our experience over the past 30 years has shown this figure to average about 20%.

Acquisition Cost Reductions

There are other costs that can be improved as a result of data management of purchased materials and simplication. They result from buying more of fewer things. The saving/cost improvements manifest themselves two ways:

Larger quantities per order means more favorable pricing breaks. You can expect an overall improvement of from 1 to 2% of the amount of annual purchases of covered items.

Purchasing expenses are reduced due to fewer purchasing transactions (i.e., the expense involved to requisition, buy, receive, pay, and account-reconcile the procured item).

The cost to execute a purchasing transaction will usually fall within a range from $15 to as high as $40 each. This has been our experience in the 30 or so years that we have been collecting these figures. One of the periodicals that specializes in reporting purchasing activities makes an annual survey of these costs. Their figures can be used as an alternative. They usually closely approximate our findings. The modal cost-figure value from our studies is approximately $22 per transaction.

If you wish to verify these figures and/or calculate your own, this can be done by determining the total salaries spent for the purchasing department, receiving (other than inspection), and accounts payable; then dividing the total by the total number of purchase orders issued. To do this, count the purchase orders issued for:

Raw materials used in product and supporting shop supplies.

Commerical commodities used in product and supporting shop supplies.

Examine in detail 3 months' purchase order output (e.g., February, April, and October of the preceding fiscal year) and record data such as:

The same item bought under different part numbers.

Purchase orders issued to the same vendor on the same day, or within 5 working days.

Different prices paid for the same item.

The average number of items per purchase order.

The increase in order quantities per order with a variety simplification of X percent.

Unit price differentials for purchase orders issued for the same items where there is a quantity difference.

You will then be able to synthesize the reduction figures in the number of purchasing transactions annually, and the overall unit costs due to buying greater quantities of fewer items.

The difficulty in the preceding analysis is in the estimate of the variety simplification potential. First time out, it can be a problem. We suggest that you calculate the increase in order quantities using a simplification range of 25–40%. That is, calculate the order quantity increase at 25% and at 40%, and you can be 90% certain that your actual simplification rate will be within these bounds.

You may use the same approach when calculating the reduction potential in the number of purchasing transactions. A good rule of thumb is that there is a 3:1 ratio between the simplification rate and the reduction in procurements (e.g., 30% reduction in variety will reduce the number of purchase orders issued by approximately 10% of the total).

For most of you, this will be a one-time event, but it can also be a lengthy and tedious study. So you may want to use a reasonable estimate and let it go at that. If you do, you can use the following:

36% reduction in variety
12% reduction in number of purchasing transactions (use 6% if you aren't controlling staff labor too well!)
$22 per purchasing transaction
1% reduction in unit costs overall
20% cost to carry inventory for 1 year

Summary of Cost Improvement Potential and Reduction in Inventory for Nonproprietary In-Product Materials

To simulate the impact of classification, codification, and simplification of such materials, we will assume hypothetical figures as follows:

Investment in inventory, $5,000,000
Annual cost to carry inventory, 20% of value per annum
Variety reduction from simplification, 36%
Purchase transactions, 10,000/year
Cost to execute a purchase transaction, $20 each

Total value of procurement annually, $10,000,000
1% lower unit costs

The savings calculations would be as follows:

One-time reduction in inventory,
$5,000,000 × 12.0% = $600,000
(from Figure 2.1, 36% variety reduction
= 12% capital reduction)
Reduction in inventory portion cost to carry,
$600,000 × 20% = $120,000 per year
Reduction in purchasing transactions costs,

$$\frac{\frac{36\%}{3} \times 10{,}000 \text{ p.o./year} \times \$22 \text{ each}}{2} = \$13{,}200 \text{ per year}$$

Reduction in costs of total annual procurements,
1% × $10,000,000 = $100,000 per year

If we presume that it will require 1 year to clear the superfluous variety, the financial benefits for the first full year after implementation will total $833,240. The annual return thereafter will be $233,240 per year.

The inventory turn rate prior to the project was $10,000,000 procurements divided by $5,000,000 inventory equals two times per year. The turn rate after year 2 will be $9,900,000 divided by $4,400,000 equals 2.25 times per year. This is a 12½% increase in the rate of rotation of the inventory investment.

It would be pointless to wax eloquent over simulated figures. It is only meaningful when your own figures are used and the picture reflected is your image, not a hypothetical collection of figures. The point of this exercise is to guide you in doing just that. There are, however, several qualifications to the realization of these benefits portrayed here. They are as follows:

1. There must be ongoing control of variety to prevent the proliferation from continuing as before. Without this control, or variety control link in the management chain, the old, bad habits will quickly reassert themselves.
2. There must be dedication to disposing of the excessive variety of items by methods other than simply scrapping the items affected. These include pooling for use-up, transferral of items to service status, sales back to vendors, and/or trading nonpreferred for preferred items. The numbers involved usually attract the attention of

management, and approval to organize to carry out this disposition project is not difficult to obtain.

2.4.2 Managing Data Variety for Proprietary Parts and Assemblies

In the design and manufacturing of in-production parts and assemblies, there are other very lucrative sources of cost improvement resulting from simplification of data. The savings are realized by avoiding the redundant work resulting from out-of-control data—data that are, for the most part, irretrievable.

You may have tried tracking down data or information from among tens of thousands of nonclassified records (drawings, data sheets, specifications, or test results). If so, you know the time taken and/or frustrations encountered. Work-sampling studies of a number of different firms by several investigating agencies reveal that up to 15% of total available design and drafting hours is spent trying to research information which is known to exist but which, because it is not in logical order, is not easily retrieved. The result is, of course, to create another data item—a new part drawing and part number.

The Part Creation Cycle

When a new part is to be designed, the following sequence of events will occur:

1. The part concept will be researched from data such as layouts, assemblies, and/or other part designs to find the state of the art for similar applications.
2. The part is then physically designed, engineered, and placed pictorially upon some medium (onto a piece of paper—the simplest procedure) or onto a visual display unit (interacting with a computer—the most complex). Supplemental documentation will then be prepared, identified, and included on a bill(s) of materials.
3. A prototype of the part or assembly may be built in the model shop and, perhaps, life-tested for history and/or destructed for performance limitations.
4. The part design is then signed off and released (with or without a design/manufacturing conference).
5. A manufacturing engineer will then determine the method of manufacture (i.e., the sequence of operations to make the part) and what tools and gages will be needed.

6. A tool designer designs the tools and gages required.
7. Toolmakers then make the tools and gages (or they are made and bought from an outside shop).
8. Work study engineers develop the working methods (motion sequences) necessary to operate the machine(s), use the tools and gages, and handle the parts. They establish the times standards for setting up the machines (i.e., to install the tooling and produce the first piece, and assure its quality) and determine the each-piece time for each operation performed.
9. A production controller adds the record, a new bin card, and sets aside storage space for the new part.
10. The cost accountant determines the cost of materials per piece, the cost of setup (per lot) and each piece time, and consolidates the costs/credits by costs center, adding appropriate burden charges, and creates the record for the particular cost system used.
11. Data processors add and store in the file the part number and all relevant data pertaining to it.

Cost to Create and Prepare a New Part for Production

In a survey taken in 1969, the average cost to do all these things amounted to $1485 for each new part. With the inflation index increasing nearly 7% per year during the intervening 8 years, the figure has surpassed $2000 each, by now.

In 30 years of examining productivity in product design activity areas, we have found that from 4% (lowest ever) to 46% (highest ever) of the parts and assemblies designed were redundant. The model value of the left-skewed distribution shown in Figure 2.2 is 11.5% of all recorded experiences.

Examination reveals that the total population will have significantly more pairs, or preventable designs, than did the sample result (see Figure 2.3). This is because the twinning attribute is not apparent unless each twin is present. Figure 2.2 verifies the fact that this is virtually never the case when sampling a population of designs. It is necessary, therefore, to apply the Bayesian approximation solution for pairs sampling, developed by Levow.*

*R. Levow, Duplication rates and sampling for matching pairs. Florida Atlantic University Research Report, 1974.

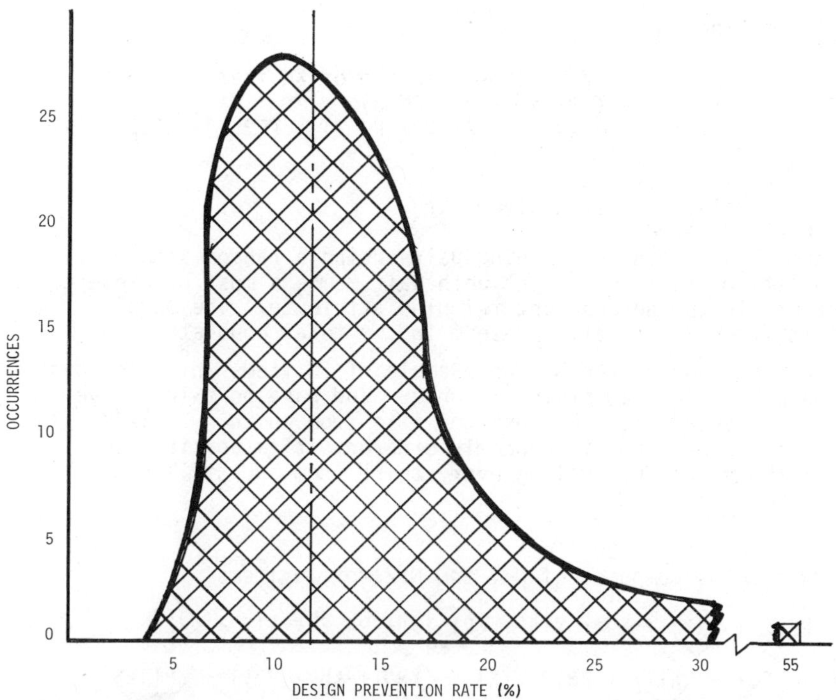

Figure 2.2 Design data redundancy prevented by classification and coding versus number of occurrences.

Calculating Savings Due to Preventing Redundant Parts

The next step is to determine the cost avoidance potential if an unnecessary new design can be prevented from entering the system. There are at least two methods to develop this figure.

Randomly select 100 parts from the sample and have each department estimate how long and/or how much it would have cost at today's costs to add the new part to the system.

An easier method is to acquire budget figures for each functional area involved, and divide by the number of new part numbers added to the system in an average year. Factor this figure by 66% to account for the fixed expense.

Pairs Sampling Problem Statement

Given: A population of N items composed of XN items without a twin and PN.1 items with twins (PN.2).

Thus, [X1 + X2 + ··· + XN] + [(P1.1 + P1.2) + (P2.1 + P2.2) + ··· + (PN.1 + PN.2)] = N

or XN + PN.1 + PN.2 = N

The total unique items, however, are

N - PN.2 = XN + PN.1

The complication in determining pairs using a random sample of a population of items is that both pair members must be present to ascertain the nonapparent twinning attribute. The sample quantity then is actually present in the total population.

This can be appreciated by the example of 30 items which follows. In the population, 20 items are unique and have no twin. Five items do have a twin. Because the file from which a 33 1/3% sample is to be drawn is randomly ordered, it is possible to select the sample by pulling every third item and still be free of a bias.

XN = no twins = 20 items
PN = twins composed of N.1 and N.2 = five items

X7 + P5.2 + X2 + X9 + X15 + P1.1 + P4.2 + X17 + X20
+ X6 + P2.1 + X11 + P3.2 + X1 + X3 + X5 + P3.1 + P1.2
+ X3 + X4 + X16 + X18 + X19 + P2.2 + X10 + P.4 + X13
+ X14 + X12 + P5.1

When the 33 1/3% sample is taken by selecting every third item, there will be one pair, i.e., P1.1 and P1.2. This is only 3%. The total population has five pairs or 16.7%.

The problem: "To determine mathematically a relationship between total population and sample results."

Figure 2.3 Pairs sampling problem statement.

Or, if you wish, use the $2000 figure; it can't be too far off. If you are of a conservation nature, use one-half of the figure.

Calculate cost improvement due to increased productivity. The staff involved in new design preparation are spending up to 18% of their available time researching data from an out-of-control data file. This figure will be reduced to less than 2% once the data base is under control. To remain conservative, and allowing for the operation of Parkinson's Second Law, use 10% reduction (if existing costs are attainable). The total these cost improvement opportunities will give an approximation figure for out-of-control data.

Other Data-Control Benefits

There are other wasted dollars that we have not dwelt upon: for example, maintenance expense inflated because of the lack of visibility of spare parts and repairs data; lost throughput because of lower-than-necessary utilization of the physical plant; higher-than-necessary tooling costs because of lack of data on existing tools, patterns, and jigs; and so on and so on.

There is one more benefit upon which it is difficult to place a value, but it is perhaps the most important of all. What is it worth to a manager to expand control into new areas lacking control, where the lack of control breeds hundreds of minor problems requiring extensive management time to resolve them? When the manager's time is spent wrestling with problems in detail, there is little time left for planning. If niggling, bothersome, annoying little fires can be stopped before they start, many problems disappear, because an orderly procedure handles them routinely. The manager is spared time to create, innovate, motivate, and establish objectives and plans for achieving these objectives. Minor problems must still be resolved, of course, but the improved use of management time is of inestimable value.

2.4.3 Factors Affecting Program Costs

Conditions that affect the program time and cost include the following:

1. Quality of source data: the extent to which data must be researched before they can be considered ready for classification
2. The method of capturing data: can the capture be done by machine, or is some degree of hand transcription necessary?
3. The size of the active data population
4. The type of data classified and coded
5. The hourly charges for the personnel involved
6. Productivity of personnel

The costs of a program include many different steps, each of which can vary according to who does it—whether or not professionals are employed—and to the extent that the control program will involve the computer. It is possible only to provide broad guidelines, as the per-computer. It is possible only to provide broad guidelines, as the permutation of variables affecting costs would suggest. There are program development costs, including allowances for training users and installing the controls procedure. Maintenance costs are not included.

I. Costs per Item for Nonproprietary Materials of All Kinds

The range of costs per item will cover the nonproprietary categories of materials that are purchased from another maker's catalog or to a recognized standard.

A. Costs per Item to Develop Data Control Solely with Internal Personnel with Limited or No Prior Experience

The expected rate of production per item to identify, classify, simplify, codify, and catalog will range from as low as 6 items per person per day to perhaps as many as 12 items each. It must be remembered that this is a one-time experience, and fraught with pitfalls for the novice.

The hourly charge will vary considerable from program to program. An expected range of $10–$15 per person per hour is not unreasonable. On this basis, the lowest labor cost per item (not counting software development) would be $6.67 each. The highest cost could reach $20 per item coded. From experience with firms who have done this themselves, a reasonable average labor cost might be $10 per item.

Software development can be expected to cost from $12,000 to $15,000. You will have to add a charge to each item to cover this. It will vary directly with the population size in your case. Using an estimated $14,000 and 28,000 items, you would add $0.50 for each item to cover the cost of program development. In 1967, an official of the U.S. Federal Stock Control System said that the average cost per item coded with a federal stock number was $7.40, give or take a few pennies. Bear in mind that this figure was for identifying each item and applying an already developed classification and code. Allowances for inflation would double that figure. To develop the code would increase the unit cost even more—by how much is anyone's guess.

B. Costs per Item Controlled with Professional Assistance to Plan, Train, and Guide the Program

Our experience with bipartite teams of, say, one professional and three or four inexperienced internals has been an average rate of production overall of 25–40 items each per day per person. The internal salaries expressed per item coded ($10—$15 range) would be as low as $2.00 for each item to as high as $4.80 each. The cost for the professional assistance will vary from program to program because it is a semivariable cost. Presuming at least 24,000 items in the population, the professional assistance costs will be approximately $2.80/item, including fees and software to catalog the data. Expect, therefore, a cost of $4.80 to $7.60 for each item controlled if using professional assistance.

II. Costs to Develop a Control for Proprietary Items of All Kinds

Parts and assemblies require approximately the same amount of time per item. Product data controls, however, vary greatly from one firm to another by virtue of a number of influencing factors. We will exclude product data—not the materials and components that comprise the product(s).

A. Controls Development Using Inexperienced People Solely

The production rate of these people will vary from 20 to 25 items per day per person. The pay rate of $10–$15 per hour per person will produce a costs that might be as low as $3.20 for each item controlled to as high as $6.00 for each item.

B. Costs per Item Controlled with Professional Assistance to Plan, Organize, Train, Guide, and Install the System

The production rate is considerably higher per person, averaging from 70 to 80 items per person per day overall. Little identification is required because drawings are used as the data base. Your internal cost will be as low as $1.00 for each item to as high as $1.71 each. Professional costs will approximate $1.84 per item, inclusive. The cost range you may expect for each item will be between $2.84 and $3.50.

III. Return-on-Investment Rate Calculation

The return on investment will probably be in the range from 75 to 400% per year. It will be up to you to determine this yardstick, using your projected savings and costs. One thing is certain—it will usually return a figure that will surpass investment guidelines by a wide margin.

You should take note of another fact. The average life of a well-designed data control system, based upon a soundly conceived, tailor-made classification and coding system, is about 25 years. It is one investment that does not deteriorate with age, but improves.

3

Classification: Principles, Methods, and Procedures

3.0 WHAT CLASSIFICATION IS AND WHAT IT ISN'T

Earlier we defined *classification* as:

> A technique to organize any related data in a logical and systematic order that groups like things together.

This differs somewhat from the definition given in *Webster's Third New International Dictionary,* which follows:

> 1. A classifying or being classified, arrangement according to some systematic division into classes or groups. 2. In biology, a system or arranging all living organisms into groups, based on some factor common to each, as structure or natural relationship; from the broadest to the narrowest, phylum (in botany, division), class, order, genus, species and variety.

Our interest is in industrial, including governmental or institutional, use of the technique. The following definition was coined for a classification adapted to these specific applications by Joseph Gombinski:

> A technique to organize specific data relating to the relevant component element(s) of a business or institution in a logical and systematical hierarchy, whereby like things are brought together by virtue of their similarities, and then separated by their essential differences.

This application of classification derives from the works of Linnaeus in the eighteenth century. Linnaeus described his classification technique in a book published in 1758 entitled *Systema Naturae*. The book presented a classification method and a standard nomenclature for all flora and fauna known to exist at the time. The nomenclature, or identifiers, are in Latin, a language not subject to change. His original work remains virtually unchanged except for the manifold expansion to include subsequent discoveries.

The Linnaean hierarchical and systematical taxonomy, showing the classification structure, is presented in Figure 3.1.

Edward Brisch, a mechanical engineer and designer, adopted the hierarchical classification of the taxonomist to aid with memory problems. In retrospect, his memory was probably no better or worse than average. What was outstanding was his sensitivity to the violation of good business practices that resulted from incurring waste needlessly. That is what nagged him as a manager—perhaps, more so, than it would most of us.

Brisch was impeccable in dress and manner. His sense of order was offended when he could not find needed design data in the system—data that were not retrievable because of lack of procedure. We have all known this frustration, but he did something about it.

It was also very annoying to him that practically no standardization was possible for most designed parts that were not commercially available. At a meeting with several members of the British Standards Institute, he said in a fit of pique: "We have standards for farthing (¼ penny), fixings and fastenings, whilst hundred quid parts (£100 sterling) are not standardized because they cannot be seen."

Data needed to aid in the design conception, or simply to review the state of the art, were in his design files but required many hours to search out. To Brisch, this simply could not be tolerated. Under great pressure to complete the suspension designs for a new British tank, both his schedule and budget were being affected unfavorably. He turned to the tool that scientists have used for centuries to organize data for research—Linnaean classification—but with a unique twist.

Figure 3.1 displays the structure of a classification for a specific application. The branching, bushlike figure is aptly named the Tree of Prophyrios, Greek for Christmas tree. It is a useful means for branching and dividing populations into defined categories. It is used to solve logic relationships and to display the parameters of commonality (or exclu-

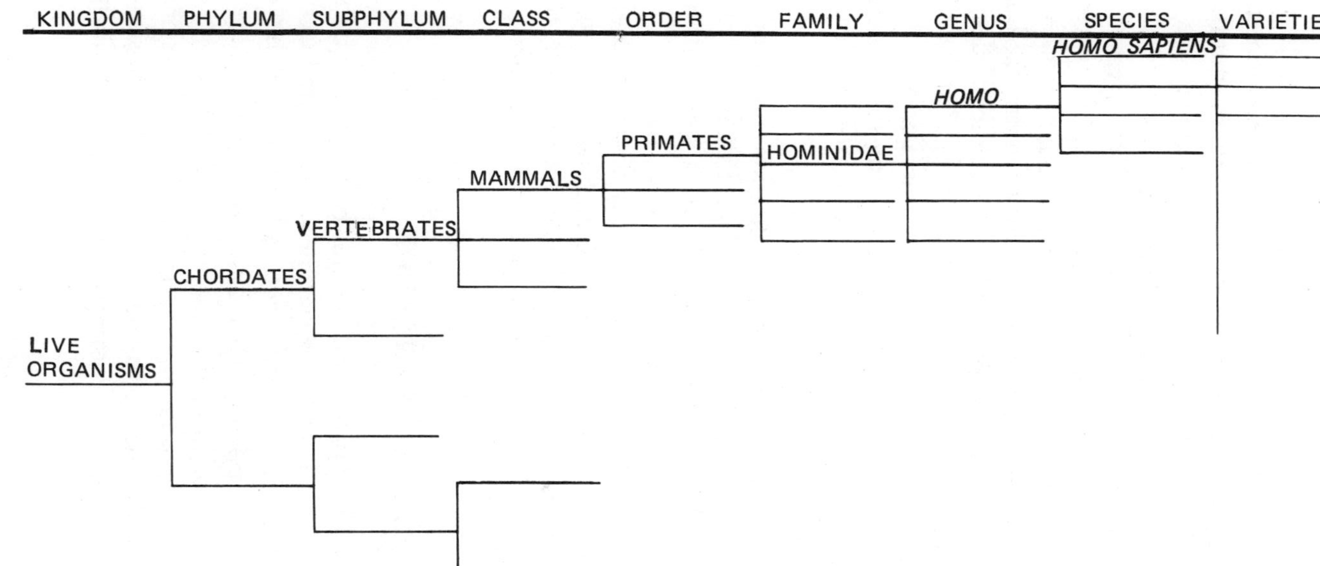

Figure 3.1 Linnaean taxonomy of human beings: an example of the Tree of Prophyrios.

sivity) which hallmark a sound classification. The Linnaean taxonomic structure is fixed and rigid. At the time of its development, the total population of things it was to embrace was unknown but thought to be finite. The rigidity of the structure does not cater to the distribution necessary for retrieval. That is a fixed and rigid structure universally applied cannot assure manageable-size data families necessary for fast and accurate retrieval. In the taxonomic classification, one species may be of a single variety or 10,000 variations of the species.

Brisch developed a structure that may appear fixed and rigid as does the taxonomic structure but, in fact, is completely flexible, to cater for the population distribution at the several levels of the classification. He could not use the rigid structure of the taxonomic classification because that would have required standardizing the nomenclature and populations of fixed parameters. This is, of course, not the case in the industrial environment because:

Most items are named to reflect their application, not what they are. For example, we have a list of 109 different names for thin, metallic parts of single outside diameter with a single inside hole diameter — one name for which is "washer, flat."

The names of parts do not cater to quantity distribution. The effect of this is clearly shown in Figure 3.2. Here you can see from a feasibility study made for a truck manufacturer why this classification did not prove effective: 74% of the items are in 17% of the families.

3.1 PRINCIPLES OF CLASSIFICATION

Brisch evolved his classification on the permanent and nonchanging characteristics of the items to be classified. He adhered to the two basic principles of taxonomic classification, adding two of his own. These principles, as now updated, when adhered to will produce the desired result—a classification of any population of things that will endure for a reasonable period of time (we try to design classifications to last 25 years), distributing the population into manageable-size families of things with parameters in common. The following principles are applied:

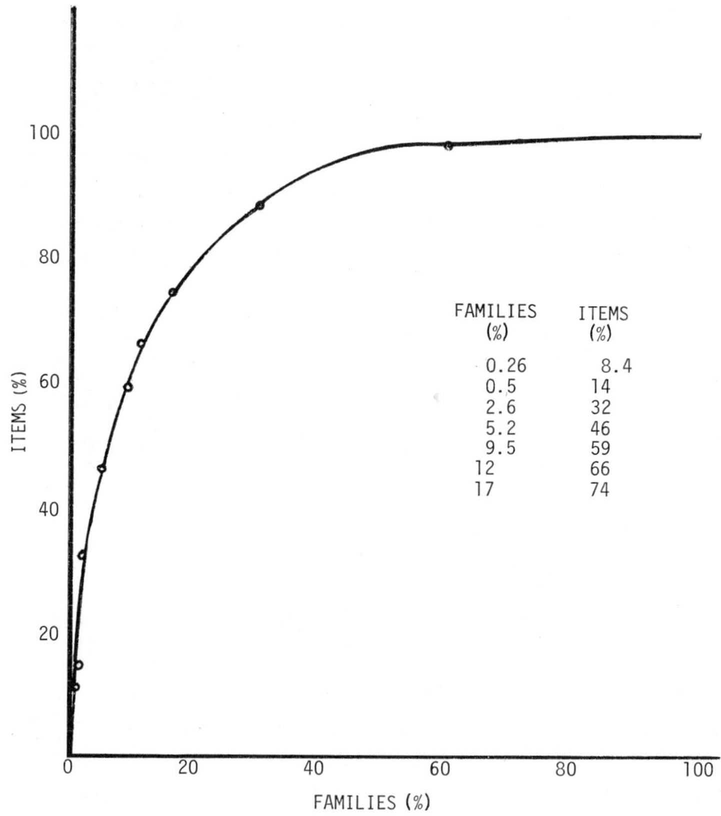

Figure 3.2 Nomenclature (noun descriptor) distribution of items per family.

1. *All-Embracing:* A classification must embrace all existing items and be able to accept necessary new items into the defined population of items. Corollary: There must be space to accommodate inclusions for the next 25 years, tailored to provide the balance.
2. *Mutually Exclusive:* A classification must be mutually exclusive, that is, include like things to bring them together while excluding unlike things, using clearly defined parameters. Corollary: There must be one place and one place only for any one item.

3. *Based on Permanent Characteristics:* A classification must be based upon visible attributes or easily confirmed permanent and unchanging characteristics. Corollary: Fortuitous, ambiguous, and nonpermanent characteristics must not be used.
4. *From User's Viewpoint:* A classification should be developed from a single point of view—that of the user and not the classifier. Corollary: Preconceived or otherwise set ideas on a solution not dictated by the facts at hand destroy the necessary overview prespective that is necessary to solve any one firm's data management needs.

You will note that each principle has a corollary. These are additions to provide additional criteria for the specific adaptation of the tool for industrial and institutional needs. Items 2 and 3 in the list are original to all classifications prior to Brisch. Items 1 and 4 have evolved over the past 30 years of industrial application of classification.

3.2 PURPOSE OF CLASSIFICATION

The user of a classification must decide for what purpose a classification is to be developed: that is, who in the organization is to be the primary beneficiary of the end product for what use?

The application of a taxonomic classification structure with the identical number of categories may suffice, but it would be a coincidence rather than a planned certainty. Figure 3.3, a raw materials classification

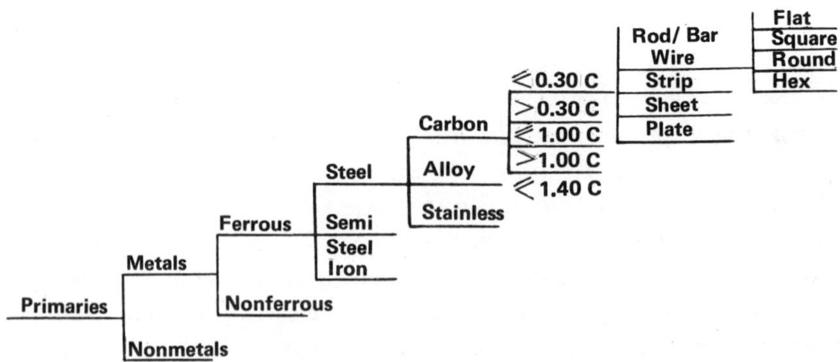

Figure 3.3 Industrial application of a taxonomic structure: classification of primary materials.

using the fixed levels of the taxonomic classification shown, demonstrates why this is so. The number of parameters shown are identical to those in Figure 3.1, but the classification does not include all of those necessary to manage an inventory. Not included are the following:

Finish
Size
Treatment
Additives o/t alloying elements

What we did here was to fit a problem to an existing solution, not design a solution for an existing problem. Brisch, on the other hand, used the tool to construct a solution for an existing problem—his problem—and resolved it to his definition of the problem at that time. It was a very simple classification, using external shape configuration and design features — features for parts that had multiple function and application potential. Parts that had single function application or were specified for a single use were designated as "specified items." He used their unique names, relating to the function they served, to organize them. Such parts as *track link pins, bogey idler wheel, gear, spring, crankshaft, nameplate,* and the like were so designated.

As coarse and unsophisticated as it may be in light of present practice, the classification order was used to arrange the files into families of similar parts of from 20 to 300 drawings each. This greatly reduced search time, practically eliminated redundant design work, and increased the productivity of his group. It was especially useful in the design concept stage, where successful solutions to design problems could be reviewed as well as those that were not successful. Brisch always maintained that it was more important to know what didn't work (unsatisfactory designs), and why, than to know what did work and not know why.

Gombinski later was to create the systematic approach that made Brisch's ideas practical, feasible, and teachable.

3.3 DEFINING THE SCOPE OF A CLASSIFICATION

In establishing classifications, it is essential to define the scope of the population to be embraced. Principle 1 says that the classification must be all-embracing but at the same time recognize a useful but finite life.

Some of the first classifications have now reached the limits of their design life. One example involves an English firm that manufactures numbering machines. The data management system was installed in 1950. It was 26 years old in 1976 and the classification was starting to show signs of age. Actually, the classification was in trouble only because the system of measurement was identified. With metrication about to make obsolete the Imperial system of weights and measures in the United Kingdom, it was obvious that some revision was going to be necessary. Today, this would not happen, because the classifications should not recognize systems of measure. This is now handled in another way, as you will see in Chapter 4.

References to measure were removed and space provided for future needs as anticipated by the long-range planning committee. The point is that 26 years earlier, when the classification was designed, the idea of full metrication in the United Kingdom was preposterous. Where metric measure was commonly encountered, the classification distinguished it from Imperial measure; this applied to about 22% of the families. But because measure was called out at all, it restricted the remainder of the classification of nonmetric categories from accepting metric items as they replaced the Imperial-measured items. At any rate, the classification was amended, with little disturbance to items already classified, coded, and part-numbered, and should last another 25 years or so.

When you seek to create a means to classify any population of data, the classification scope definition must be as clean and free of ambiguity as possible. A typical definition for a class of data is as follows:

> *Class: Primary Materials.* Those items bought in bulk to a recognized standard or industry specification to which value is added to change at least one characteristic before it becomes part of the product, is used in production as a supporting aid in manufacture, and/or is used to maintain the facility. Included are such typical items as: paint, chemicals, lubricants, cutting oils, and coolants; conducting wire, buss, cable, conduit, and insulation; strip, sheet, plate, and flat bar; rod, wire, round bar, pipe, and tube; ores, earths, and alloying elements; fabrics and tapes; solders, fluxes, and gases; adhesives, binders, and elastomers.

The definition is not intended to be universal. It is likely to differ from application to application if it is to be all-embracing for any one firm's needs. It does illustrate how a category of related items can be qualified for consideration.

The user's point of view must be reflected in the definition. In a chemical manufacturing plant, for example, maintenance requirements transcend manufacturing requirements. Productive raw materials may be few (e.g., 10 or 20 hydrocarbon feedstocks), whereas nonproduct items may be in the tens of thousands.

To be mutually exclusive in defining the category or class of data is not as simple as it first appears. Especially is this true for, say, extrusions. Although these are considered a raw material, in that they are produced to random or specified lengths comparable to other mill products, they are generally of a crosssection created by or for the user to a specific design. They are, nonetheless, considered raw materials with the standard or specification defined by the user. On the other hand, raw castings and forgings are *not* classified as raw materials but as proprietary designed parts.

Another area of seeming overlap is in commercial, nonproprietary items. These kinds of items are defined as:

> *Class: Commercial Nonproprietary Parts and Assemblies.* Those items bought to a recognized national, industrial, institutional, or company standards and/or specifications; the design control of which is exercised by other than the user. Included are such typical items as: fasteners; bearings, seals, and bushings; pumps, pipe fittings, cocks, and valves; electric motors, controllers, stators, and switches; solenoids, relays, interrupters, and coils; fuses, circuit breakers, stand-offs, insulators, and conduit fittings; hydraulic cylinders, pumps, actuators, and tanks.

For the vast majority of items, there is no real debate. For some 10–15% of the items, there is a question of overlap with other classes of materials. This can be dramatized by an actual example.

Direct-current drives for processing machinery require literally pages of specifications when purchasing the product from the maker. Although the catalogs call out the product overall performance, the options offered require a considerable number of decisions by the buyer. If the options are within the purview of the industry recommendations (e.g., NEMA standards for electrical apparatus), this class of materials most immediately defined can include them. If, however, mounting-hole locations and sizes, terminal locations and connectors, shaft length and diameter, driving configuration (keys, splines, flats, and the like) are to be specified by the user, there is every reason to question whether the end product is a designed assembly or a commercial nonproprietary part or assembly.

The experienced classifier will deal with these items using the inclusion rule—the majority attracts the minority—and do it consistently. The inexperienced classifier will vacillate first one way and then another. So some common sense must be applied, no matter how far you go in attempting to make a precise definition.

This leads to another point regarding defintions. In classification work, the binary form of decision—yes, it is; or no, it is not—is desired and should be adhered to. Consequently, the text of the definitions you use for a class of materials should not require a shopping-list method of use. That is, do not require the person using the classification to go through a table-look-up search. Although examples are useful to aid the user in selecting the appropriate class of items, they are not intended to substitute for the classification.

3.4 LOGIC AND CLASSIFICATION

Mutually exclusive parameters of primary, visible attributes are necessary to speed use and reliability. How would you classify books in a library? One way could be by their color, but having all puce books together would mean nothing to someone who wanted a certain fact contained in a book. And certainly, a classification by book shape would be an equal waste of time and useless as a means of retrieving data.

Yet surprisingly, people develop classifications based upon equally meaningless attributes, such as "where-used". Now, it is essential to know where something is used, but that's what a bill of materials and a where-used file does. An experience with a firm making writing instruments illustrates this point. Where-used is neither a permanent characteristic nor a mutually exclusive characteristic. The firm in question did not use a bill-of-materials method of displaying its product components and materials. For that matter, neither did they use process records to list the operations to make the parts. Rather, they depended upon an elaborate part numbering system to try to accommodate both system deficiencies. The system was about to become exhausted, so they wanted to revise it. They asked that each material and part be classified by where it was used, i.e., model number, and how it was processed.

The classification specialist refused to undertake the project with these constraints because two of the four classification principles (numbers 2 and 3) were being compromised. Principle 4 (the user's point of view) was not allowed to compromise the other three principles.

Another example of this is in developing a chart-of-accounts classification, where the user wanted to insert seller's identification, region, and district as they related to statement reference for selling expenses and income accounting. This meant changing the point of view from the primary requirement of finance and control to one of marketing. It is not that these data are not needed. The point is that the classification cannot substitute for deficient systems. There are other, more efficient means of capturing and displaying these needed data than by the classification.

3.5 GETTING THE BASICS

How, then, does one establish an all-embracing, mutually exclusive, classification of the permanent characteristics of a population of items while maintaining a single point of view of the user? Easy question, but it requires a lot of answering. Take a deck of playing cards and place them face down upon a tabletop after thorough shuffling. The population is finite (52) and known—13 cards, ace through king, in four suits: hearts, spades, diamonds, and clubs. If you were to develop a classification system for the cards for, say, a bridge game, you could select a 13-card sample at random. Then, take the 13-card sample and examine it as a bridge hand. You would arrange the cards first by suit—spades, then hearts, diamonds, and clubs—the bridge player's order. Then you would arrange the cards by value—highest to lowest order. A classifier would use the Tree of Prophyrios for classifying 52 playing cards for bridge, as shown in Figure 3.4.

When you set up the 13-card sample, this is what you did:

1. You *identified* each of the 13 items (cards).
2. You *classified* each of the items by its suit.
3. You *separated* each of the *varieties* in each suit by the essential differences (highest to lowest value).

You can do this more of less automatically if you are a bridge player and it has become a well-practiced pattern—a habit, if you will. The same

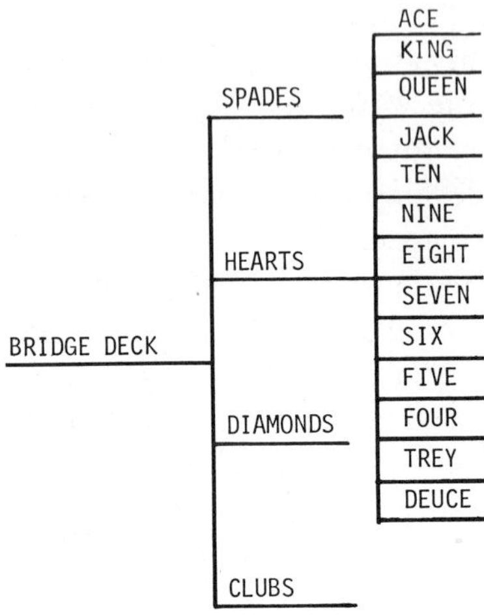

Figure 3.4 Classification order for a bridge hand.

now-classified hand would be useless to the user, however, if the game to be played were gin rummy and not bridge. The number of cards is too great and the gin player's order can be by card value and/or by suit and is arranged differently, according to the population distribution of each hand.

There are three valuable points to be made in this very simple analogy.

1. The user's point of view must be reflected in the classification.
2. The population determines the classification, not vice versa.
3. The classification varies from user to user and population to population.

In the United Kingdom, some schools use classification as a means of teaching logic. An educational tool called Attribute Logic Blocks is one device that can be used for this purpose. In the instruction literature that accompanies the blocks, it is explained that logical thinking and decision making are the keystones to a child's early progress in mathematics and science. Attribute Blocks are the most widely accepted means of introducing logic and decision problems through classification of the blocks. The permutation of the five shapes, times the three colors, times the two sizes, times the two thicknesses is 60.

In one of the training exercises, the child is asked to arrange the blocks in any order that he or she chooses. The order usually found in first-form children is:

Color
Shape
Size
Thickness (if recognized)

The children are asked if they can tell why they chose the order they did. Then they are asked to group the blocks in a different order.

At this point, we can use the card and block analogies to point out some similarities as they relate to classification. In both of these simple examples, there are these similarities in what otherwise appear to be diverse populations:

Each of the two populations of data items is fixed and finite (52 and 60 items, respectively).

Each has an equally balanced distribution of its constituent parameters (4 suits and 13 cards of unique value each; and 5 shapes, 3 colors, 2 sizes, and 2 thicknesses, respectively).

Each item in the populations is fully identified immediately upon seeing it.

The classification options are relatively few according to the user's needs for the respective classifications.

You may wish to corroborate these statements by reexamining the bridge-hand classification shown in Figure 3.4 and comparing it with the three Attribute Block classifications in Figures 3.5 to 3.7.

It is interesting to note that the following facts appear in the three parameter-order arrangements in the Attribute Block classifications:

According to the parameter order chosen, note the variety displayed in the last parameter chosen. In Figure 3.5 there are two items, in Figure 3.6 there are three items, and in Figure 3.7 there are five items.

According to the parameter order chosen, the number of families vary. In Figure 3.5 there are 30 families, in Figure 3.6 there are 20 families, and in Figure 3.7 there are 12 families.

According to the parameter order chosen, the respective classifications bring similar things together but are different from classification to classification.

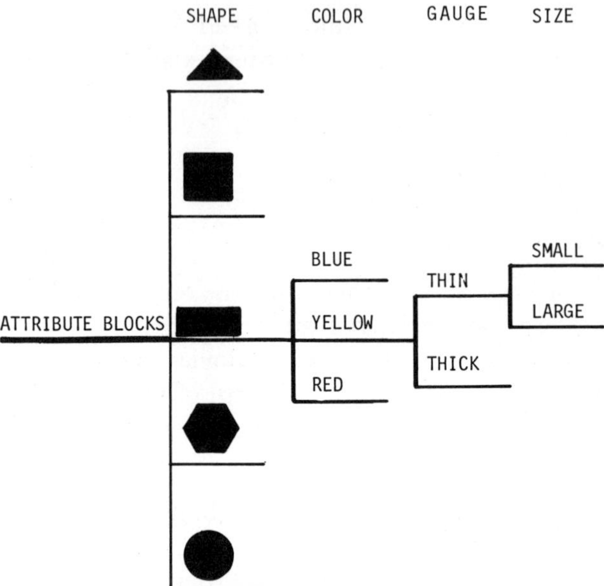

Figure 3.5 Attribute classification order by shape, color, gage, and size: 30 families.

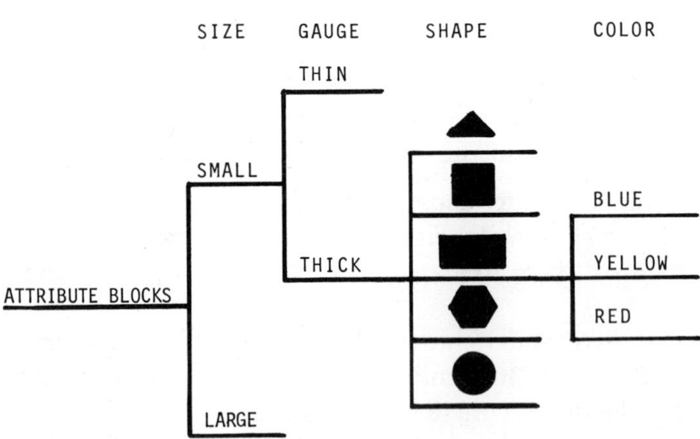

Figure 3.6 Attribute classification order by size, gage, shape, and color: 20 families.

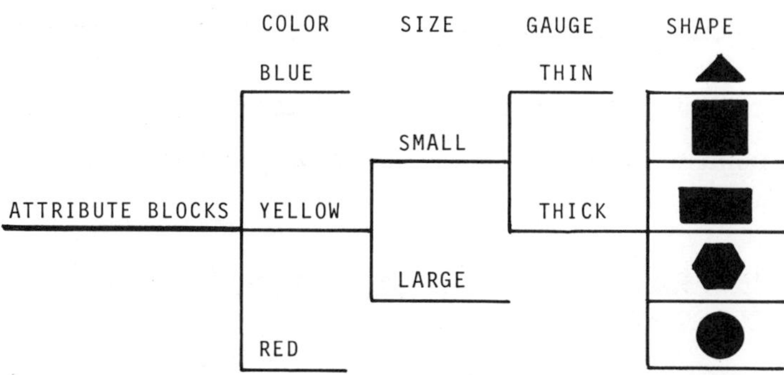

Figure 3.7 Attribute classification order by color, size, gage, and shape: 12 families.

In Figure 3.5 the classification brings together blocks of the same shape, color, and gage blocks, and the variety of the two are the essential differences in their sizes. In Figure 3.6 the classification brings together the same size, thickness, and shape of blocks, separating them in a variety of colors. In Figure 3.7 the classification brings together the same color, size, and thickness of blocks, separating them by variety of shapes.

These simulation exercises are most revealing. They confirm the four principles that prove the soundness of a good classification – bringing like things together by virtue of their similarities and then separating them by their essential differences. It can also be concluded that one should use the fewest number of parameters necessary to distribute the population in reasonably well balanced families of similar data items as dictated by the principal user's requirements.

To apply the principles of classification to larger and more diverse populations requires that certain procedures be followed. In the preceding simple examples, the entire population was easily visible and identified. In the real world of data in various degrees of disclosure and identification, procedural means are necessary to overcome this first barrier to classification of the data:

1. Define the needs of the principal users of the classification.
2. Define the population of data that is to be embraced by the classification.

3. Capture the raw data and connect it to a uniform media format for handling.
4. Identify the raw data to enable it to be classified.

At this point we interrupt the classification development program to discuss data identification.

3.6 CAPTURING RAW DATA AND IDENTIFYING THEM FOR CLASSIFICATION

According to the category of data involved, identification will vary considerably, as will the form of the data and where they are stored. For example, to classify data elements and data names in order to catalog these items requires an analysis of computer-generated reports and their supporting programs. In one such program, we found 81 data elements and/or data names for date when only two were required for the two types used (Julian and Gregorian).

On the other hand, the data may be completely identified and easily convertible to a standard format for handling. A good example of these kinds of data are engineering designs or engineering drawings. Examine Figures 3.8 through 3.10. These are extracts from a file of 24,000 plus items, reduced and copied on 5 × 8 in. paper. Figure 3.8 is a simple round part. It has six readily visible attributes, as follows:

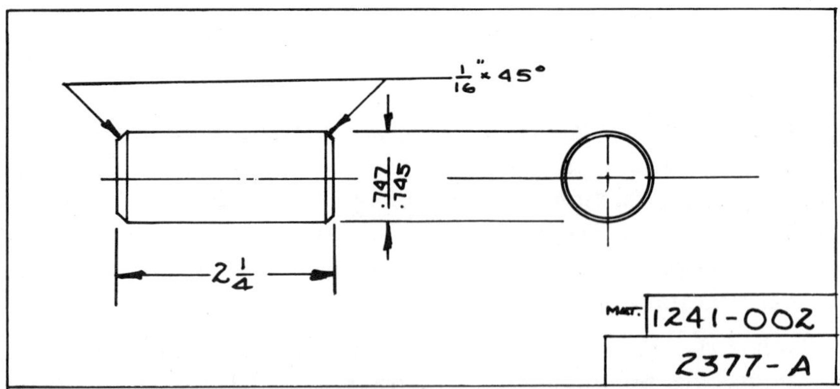

Figure 3.8 Round piece part drawing: single outside diameter.

Material from which it is made
Shape
Diameter size
Length overall
Design feature—chamfers
Number of chamfers

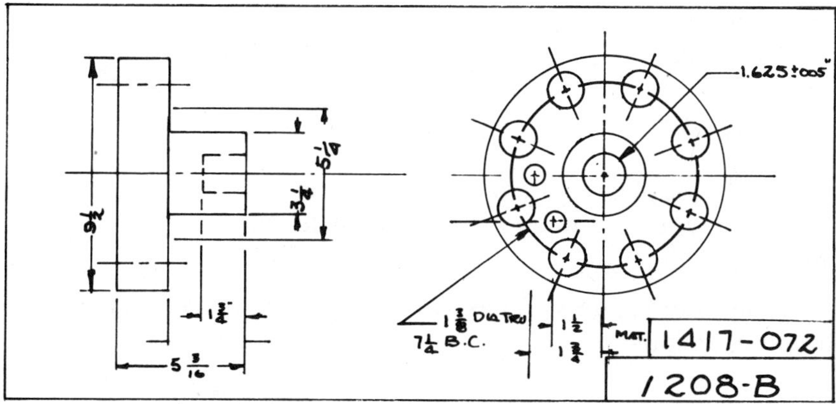

Figure 3.9 Round piece part drawing: two outside diameters.

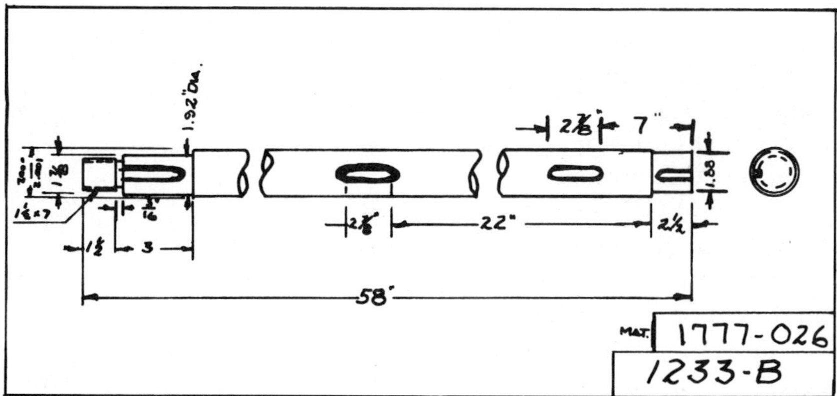

Figure 3.10 Round piece part drawing: more than two outside diameters.

By contrast, the drawing of the part in Figure 3.9 has 15 readily apparent parametric attributes that distinguish it:

Material
Shape of exterior
Number of OD portions
Location of OD portion
Size of major OD
Length of major OD
Length of part overall
Shape of interior
Centerhole depth
Number of holes on bolt circle
Bolt circle diameter
Hole diameter
Number of longitudinal holes o/t on bolt circle
Location of longitudinal holes o/t on bolt circle
Size of minor OD

The part shown in Figure 3.10 has 35 readily visible physical parametric attributes, as follows:

Material
Shape
Length overall
Size of maximum OD
Maximum OD location
Second largest OD size
Third largest OD size
Fourth largest OD size
Fifth largest OD size
Number of OD portions
Length of first OD portion
Length of second OD portion
Length of third OD portion
Length of fourth OD portion
Position of major OD portion
Location of first OD portion from base line of dimension reference point
Location of second OD portion from base line of dimension reference point

Location of third OD portion from base line of dimension reference point
Location of fourth OD portion from base line of dimension reference point
Location of fifth OD portion from base line of dimension reference point
Presence of keyways
Type of keyways
Number of keyways
Location of first keyway from base line of dimension reference point
Location of second keyway from base line of dimension reference point
Location of third keyway from base line of dimension reference point
Location of fourth keyway from base line of dimension reference point
Length of first keyway
Length of second keyway
Length of third keyway
Length of fourth keyway
Presence of threads
Type of thread
Location of thread
Length of thread

3.7 ANALYZING THE DATA POPULATION

Developing an all-embracing, mutually exclusive, visible attribute classification from a single point of view is done by preparing a preliminary analysis classification. This is done by sampling the population.

If the parts in Figures 3.8 through 3.10 are considered typical of the entire population (which they are not) and the distribution about equal, say 8000 of each of the three kinds, we could use these three items as the sample. A family size of 8000 items is obviously too great, but it can be used in this exercise. If the principal user is to be product design, it is necessary to have no more than 40 item varieties in each family. This would require approximately 600 families to embrace the 24,000 items and produce an average of 40 varieties in each family. An analysis classification of two of the three sample items to accommodate this is shown in Figure 3.11.

Note that the simple part in Figure 3.8, with few parameters, can

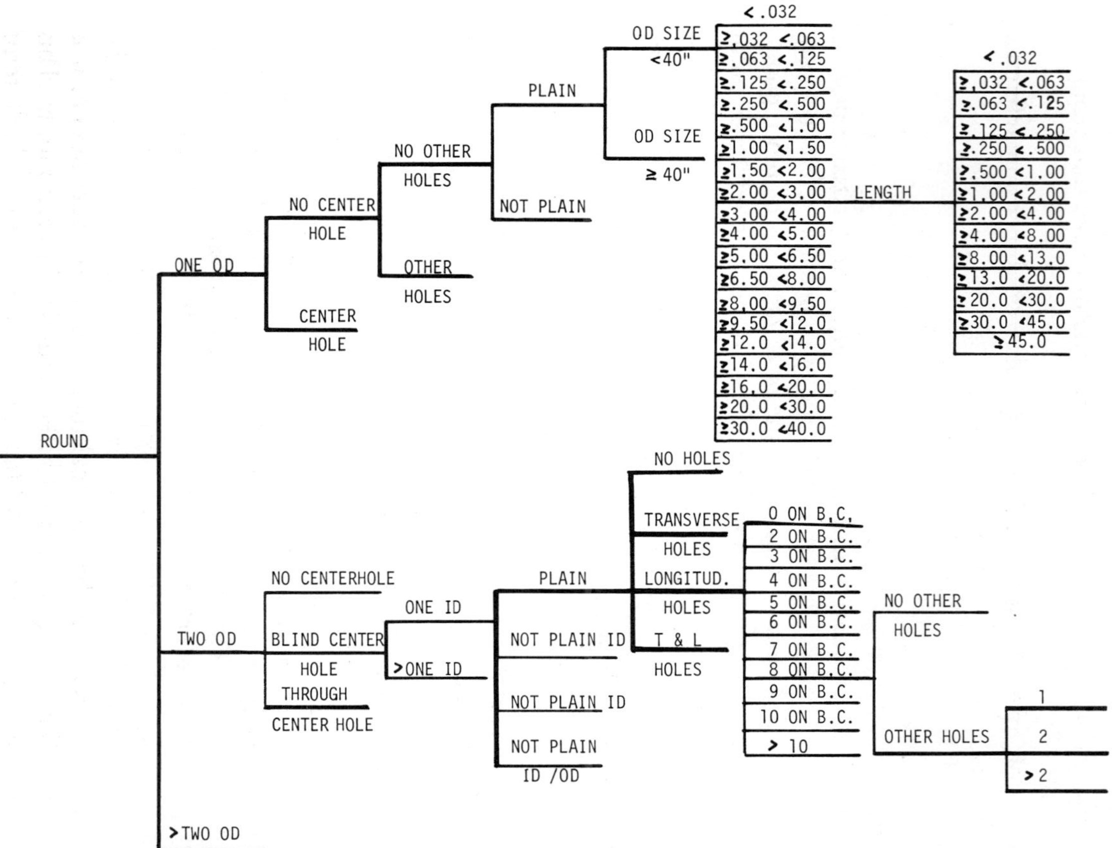

Figure 3.11 Classification of the single piece parts shown in Figures 3.8 and 3.9.

be only split on size, because material is coded in its own right. Note also that although the permutation of the diameter-size split times the length-size split produces the desired number of families, there is no guarantee, without having first analyzed the 8000 items, that the distribution of variety will produce the desired family size. And this is a very, very important observation. Although mentioned in passing earlier, it is strongly emphasized here and is not to be forgotten.

Why is this so important? Suppose that, after establishing the preliminary size splits that were made, we find that 1000 of the 8000 parts are in one family and 800 in another family. The classification must be restructured to overcome this problem, as there are too many parts in one family. What this proves is that tailormaking the classification is an absolute. It is essential to provide the desired service to the user.

Conversely, what this says is that preconceived fixed classifications simply cannot fit in the universe with any assurance that any one user's best interests will be served.

This is further borne out by an examination of the classification of the two-diameter part in Figure 3.11. The part was shown in figure 3.9. As you can see, five parameters will divide the population into 192 families. If the balance is correct, there would be an average number of items per family of 41 or 42 items each.

The parameters chosen differ from those of the single-diameter part. You must remember that we are bringing together like things and it is important to carry the classification far enough to accomplish this objective. If we were to carry the classification one level further, the permutation could conceivably produce more than 1920 families. One more split would create 3840 families and the last split, 10,860 families, or less than one item per family as the classification is presently structured.

It is fairly apparent, therefore, that a classification must be tailored to the situation. The classifier's prejudgments on other than the facts revealed by handling the data and evaluating it will be false and misleading, producing a virtually useless tool for the long term.

The steps covered up to this point in developing a classification are as follows:

1. Determine the purpose or end use of the classification (i.e., who is to use it and for what purpose).

2. Define the scope of the classification (i.e., the general class of data in the population to be embraced).
3. Capture the data population you expect to embrace.
4. Analyze the population to determine its composition. To do this, take a sample of data randomly and develop a preliminary classification to use in the analysis.
5. Quantify the distribution of the sample.
6. Determine the size of families to provide sufficient variety for the user to chose from, but not so many as to retard the selection process.
7. Synthesize the classification to provide the number of families that will distribute the population as you have planned.

3.8 SYNTHESIZING THE CLASSIFICATION STRUCTURE

It would be pointless to assume further conditions and change the classification without actual data with which to work. But we must dwell further on the synthesis step. There is no one right way to do this step. Ten trained people, using an appropriate and standardized glossary of terms such as shown in Figure 3.12, will produce a somewhat similar, but not identical result. Each will be workable and efficiently divide the population as desired. But the numbers of parameters, the order of parameters, the end user, and the objective must be interpreted in exactly the same way to produce a similar product. It goes without saying that the four principles must also be adhered to exactly.

The tailored classification, therefore, can cater to the tremendous variety of choice and produce a useful end product. If there is one thing that has become apparent over the preceding 30 years, it is that classification is an art rather than an exact science. That two classification experts might produce classifications that differ somewhat from each other is not so important as is the fact that both classifications can produce effective solutions to the data control and management problem(s).

For this reason, if for no other, when supervising the activities of specialists in this work, care has to be taken to ensure that criticism is limited to the adherence to basic principles. It is very easy to be an art critic but difficult to be an artist.

Classification is therefore considered to be a pseudoscience whereby the end result satisfies the objectives established for it even though reproduction by another specialist may vary from the original and still satisfy the objectives set for the original.

As useful as classification is, it becomes efficient as a data-control device only when it is married to a coding system, as you will see in Chapter 4. Some classifications are computerized. They are accessed by key words used in the classification to track to a family of data. This is one way to use a classification without a code. However, there is a cost attendant with this procedure. The Attribute Block classification will provide a useful example. The classification represented by Figure 3.6 had a parameter order as follows:

Size
Thickness (gage)
Shape
Color

After the computer file containing these data items and accompanying classification are stored in memory, an Operator (O) at a terminal communicates with the Computer (C) in a dialogue to find a thick, small, yellow, hexagonal part. One such method (bipolar) follows:

C: Is the part thin?
O: No.
C: Is the part small?
O: Yes.
C: Is the part triangular?
O: No.
C: Is the part square?
O: No.
C: Is the part rectangular?
O: No.
C: Is the part hexagonal?
O: Yes.
C: Is the part blue?
O: No.
C: Is the part yellow?
O: Yes.

To summarize this chapter in the fewest words possible, memorize the principles of classification and remember to establish the objective the classification is to satisfy. Then, remember to tailor the classification

TERM	DEFINITION	
BOLT CIRCLE HOLES	TWO OR MORE HOLES WHOSE AXIAL CENTER LINES ARE PARALLEL AND EQUIDISTANT FROM THE CENTER POINT.	IF PART HAS TWO BOLT CIRCLE HOLES, THEY MUST BE DIAMETRICALLY OPPOSITE. 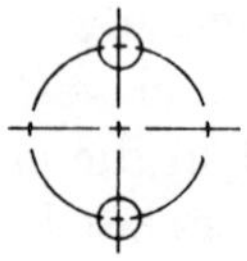
		IF PART HAS THREE OR MORE BOLT CIRCLE HOLES, THEY MUST APPEAR IN AT LEAST THREE QUANDRANTS OF A CIRCLE. HOLES MAY OR MAY NOT BE EQUI-SPACED. 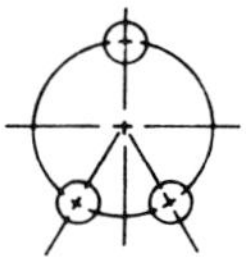
CENTERHOLE	BLIND	LONGITUDINAL HOLE(S) THAT DOES NOT PASS THROUGH THE PART AND COINCIDES WITH THE AXIAL CENTERLINE OF THE PART. 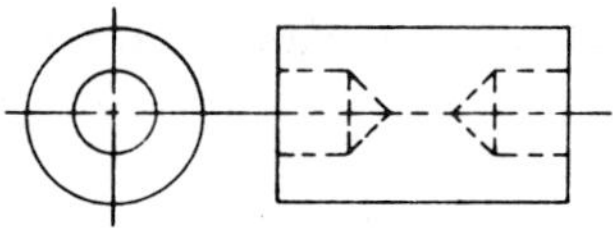
	THROUGHGOING	A LONGITUDINAL HOLE THAT PASSES THROUGH THE PART AND COINCIDES WITH THE AXIAL CENTERLINE OF THE PART. IT MAY OR MAY NOT EQUAL THE PART LENGTH. 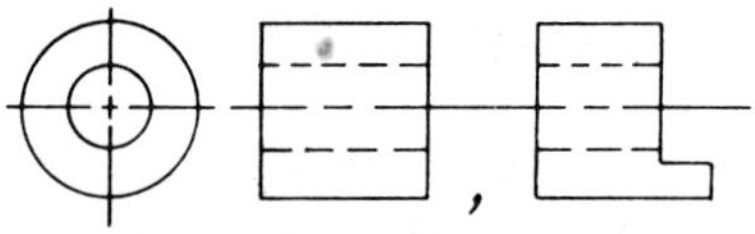
DIAMETER PORTIONS	ARE THE DIAMETERS VIEWED IN THE LENGTH VIEW. THEY ARE DETERMINED BY EACH SUCCESSIVE CHANGE IN DIAMETER. IDENTICAL DIAMETERS SEPARATED BY A DIFFERENT SIZE DIAMETER ARE CONSIDERED AS INDIVIDUAL DIAMETER PORTIONS. 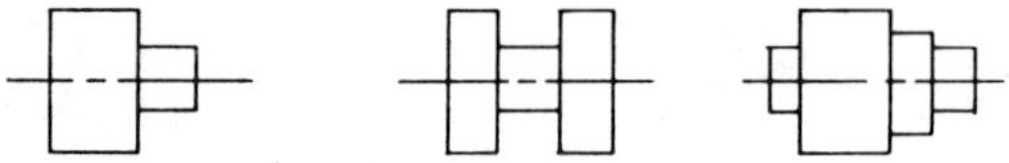 TWO THREE FOUR	

Figure 3.12 Extract from a typical glossary of terms applicable to axial parts. (Copyright © 1975 Brisch, Birn & Partners.)

TERM	DEFINITION							
	SINGLE DIAMETER		MULTIPLE DIAMETER					
			MAXIMUM ONE END		MAXIMUM BOTH ENDS		MAXIMUM BETWEEN ENDS	
	OD	ID	OD	ID	OD	ID	OD	ID
DIAMETERS								
	ALL SECTIONS SHOW SIDE VIEW ONLY; END VIEWS AND SECTIONS BEING ROUND.							
ENCLOSING PORTION INCLUDING FULLY ENCLOSED	IF THE EXTREMITIES WERE TO BE EXTENDED THEY WOULD MEET. EXTREMITIES MUST BE IN SAME PLANE. MAY BE EITHER: IN THE "ACT OF ENCLOSING" OR FULLY ENCLOSED. 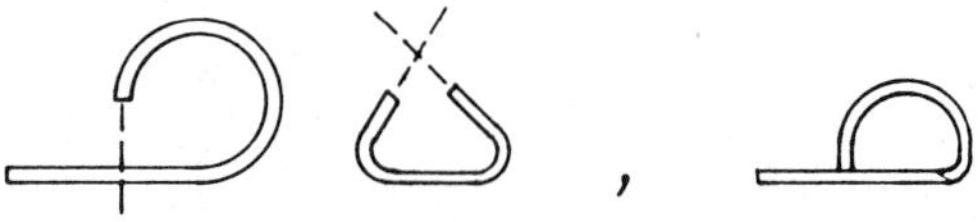							
EXTREMITIES BENT IN ONE PLANE								

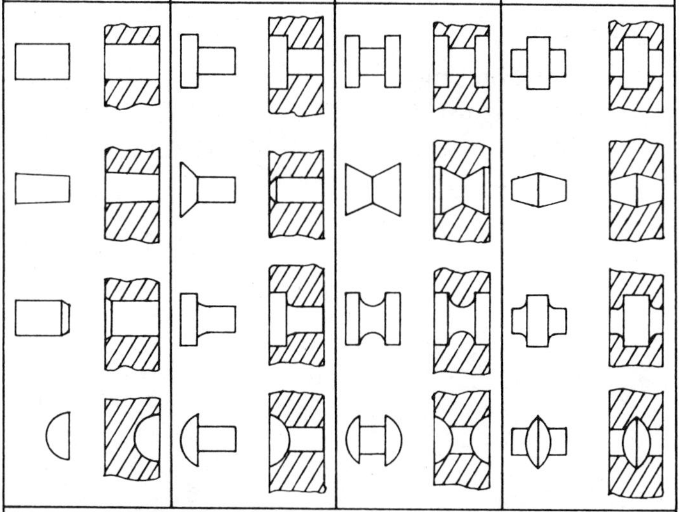

to satisfy the objectives, by restructuring the first draft analysis classification. The restructuring may require reordering the parameters chosen or revising the range of the level dealing with a parameter. Remember, you are trying to bring like things together in a logical, systematical, hierarchical order by virtue of their similarities, and then to display them by their essential differences. To do this, you must adhere to four basic principles:

1. All-embracing
2. Mutually exclusive
3. On easily identified attributes that are permanent
4. From the user's point of view

4

Coding: Principles, Methods, and Procedures

4.0 CODING DEFINED

As useful as classification is in the organization of data, alone it offers little of its potential benefits to managers. Not until the classification parameters have been codified will an efficient means to manage and handle the data (i.e., retrieve relevant information) be possible.

This may sound contradictory at first, especially so in light of the great emphasis placed upon classification in Chapter 3. It is true that classification brings like things together, but until a coding system reflecting the classification is introduced, handling and processing families of similar data is cumbersome and inefficient.

If you are like me, codes and coding conjures up secret communications from spies and secret agents, à la James bond et al. When I was growing up, hardly a week went by that either Jack Armstrong or Skye King wasn't offering a special code ring for two Wheaties box tops and 10 cents to cover postage. With this secret agent's ring, daily radio messages broadcasted in code could be decoded by the wearer, alerting us to some impending achievement by one or the other of our heroes.

For most lay persons, coding is thought of in its cryptological sense, meaning to restrict and suppress the dissemination of intelligence and/or information. In the business and industrial context, quite the reverse is intended. Properly designed industrial codes are used as a shorthand, as a means to compress information and improve its communication effectiveness and processing costs.

The usual places to research the meaning of words — dictionaries and encyclopedias — do not pecisely define the word *code* as we have come to use it. This is what *Webster's* Unabridged has to say on this subject of code, codes, and coding:

> **Code,** *n.* (fr. code, a code; L. codex, the trunk of a tree, a wooden tablet covered with wax for printing). 1. a body of laws of a nation, state, city or organization, arranged systematically for easy reference. In Roman law, THE CODE is the distinctive title which, by way of eminence, is applied to the collecion of laws and constitutions of the Roman Emperors, made by order of Justinian, containing twelve books. 2. Any accepted system of rules and regulations pertaining to a given subject, as the MEDICAL CODE, which governs the professional ethics of physicians; also a system of rules and regulations governing the conduct in particular cases, as the social code, the code of honor, etc. 3. A set of signals representing letters or numerals, used in sendng messages, as by telegraph, flags, heliograph, etc. 4. A system of secret writings in which letters, figures, etc., are arbitrarily given certain meanings. 5. The symbols used in such a system.

To code, as a verb, is defined by *Webster's* as follows:

> **Code,** *v.t.,* coded, *pt. p.p.;* coding, *p.p.r.,* to put in the form of a code; to translate into the symbols of a code.

While, for most people, codes are the stuff of Ian Fleming and James Bond, the *Ultra Secret,* and *The Man Called Intrepid,* and not of GM, IBM, Ford, and your company, there are occasions when sensitive industrial and business data are coded for security purposes, and we have had experiences of this use in coding. However, in this chapter, we will keep to the less sinister aspects and discuss coding as a means to disseminate information, not suppress it; as a means of unlocking information and data kept secret by its obscurity in files—whether magnetic, encased in steel, or on little pieces of film.

We developed our own definition for codes as used in industry as follows:

> A *code* is one or more symbols to which an arbitrarily assigned meaning and/or arrangement has been given which, when deciphered, communicates specific information or intelligence.

Our defintion of the verb *to code* is as follows:

> To translate information and/or data into the symbols of a coding

system, according to a prescribed method upon a standardized presentation format.

4.1 MANAGEMENT CODING SYSTEMS

Systems of codes for management use take several forms. Those that use the alphabet or numerals are called, logically, *alpha* or *numeric codes* (or in combination, *alphameric codes*). Other categories include special symbol codes not derived from either alphabetic or numeric sources.

4.1.1 Special Symbol Codes

Special symbol codes may be hieroglyphical. *Hieroglyphic* coding systems are best characterized by the symbolic therbligs, created by the Gilbreths, or of more recent vintage, the flow chart codes for tracking the handling of data within a system.

The hieroglyphic code symbols epitomized by the therblig codes pictorially represent the action they described. For example, the planning therblig symbol is a person with his finger at his brow, denoting a thinking pose. In the flow chart symbols, the handling of data on punch cards is represented by a symbol that looks like a punch card.

Nonhieroglyphic special symbol codes have little or no meaning implied in the shape or character of the code and the information it represents. The familiar flow process chart symbols shown in Figure 4.1 are a good example of such codes.

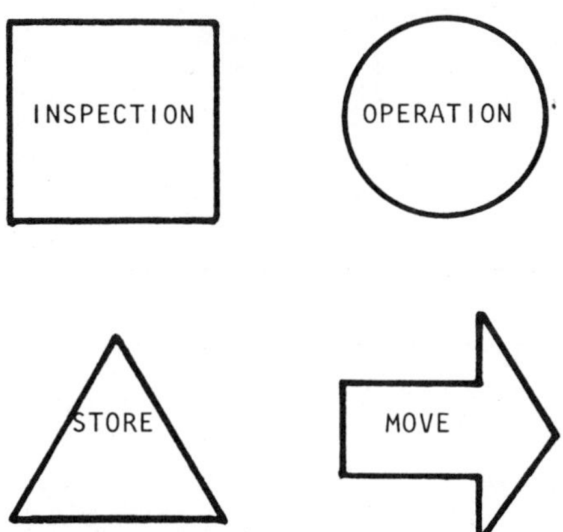

Figure 4.1 Flow process chart: nonhieroglyphic special symbols code.

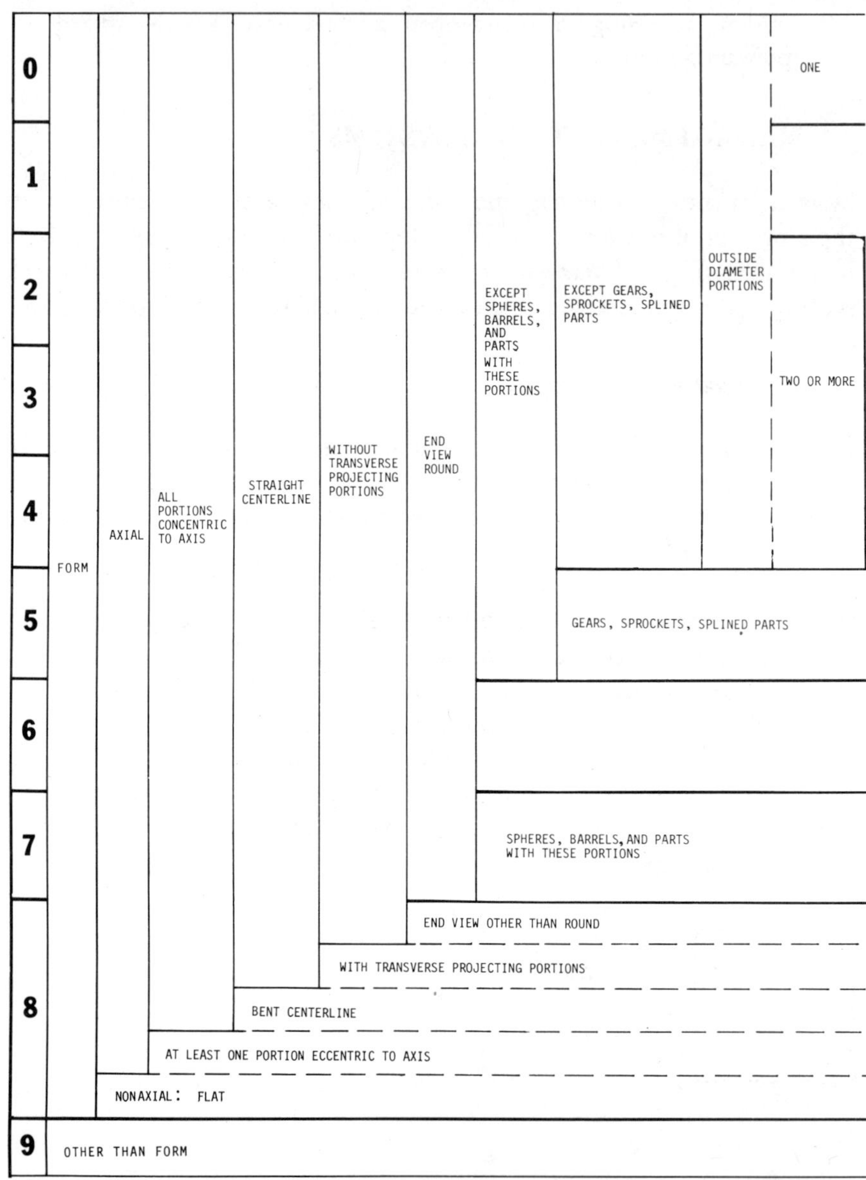

Figure 4.2 Integrated numeric code. (Copyright © Brisch, Birn & Partners.)

			0
			1
MAXIMUM OUTSIDE DIAMETER	ON ONE END ONLY		2
	BETWEEN ENDS ONLY		3
	OTHER THAN 92XX/93XX		4
			5
			6
			7
		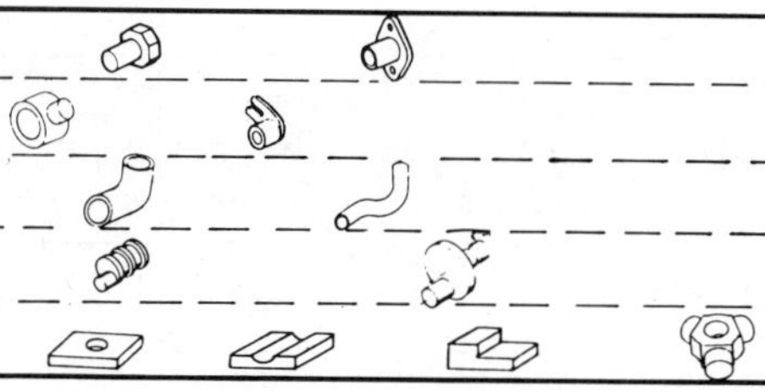	8
			9

							0	1	2
0X	OUTSIDE DIAMETERS	SUCCESSIVELY DECREASING	OUTSIDE DIAMETER PORTIONS	TWO	L ≤ D	PLAIN	WITHOUT CENTERHOLE		BLIND
1X						OTHER THAN PLAIN	WITHOUT TAPER PORTION		
2X					L > D		WITHOUT CENTERHOLE WITHOUT TAPERED OD PORTION T ≤ D	T > D	WITH TAPERED OUTSIDE DIAMETER PORTIONS
3X									
4X				THREE			WITHOUT CENTERHOLE		BLIND WITHOUT TAPERED OUTSIDE DIAMETER PORTION
5X				FOUR OR MORE			WITHOUT CENTERHOLE WITHOUT TAPERED OUTSIDE DIAMETER PORTION	WITH TAPERED OUTSIDE DIAMETER PORTION	
6X		NOT SUCCESSIVELY DECREASING	OUTSIDE DIAMETER PORTIONS	THREE			WITHOUT CENTERHOLE		
7X				FOUR OR MORE	L ≤ D		WITHOUT CENTERHOLE WITHOUT LONGITUDINAL HOLE(S)	WITH LONGITUDINAL HOLE(S)	
8X					L > D	FIVE OR LESS OUTSIDE DIAMETER PORTIONS	WITHOUT CENTERHOLE WITHOUT TAPERED OUTSIDE DIAMETER PORTION	WITH TAPERED OUTSIDE DIAMETER PORTION	
9X						SIX OR MORE OUTSIDE DIAMETER PORTIONS	WITHOUT THREADS ON OUTSIDE DIAMETER WITHOUT CENTERHOLE		WITH CENTERHOLE
							0	1	2

Figure 4.2 (continued)

In all these coding systems, significance was assigned. The significance may be arbitrary or directly interpretive (e.g., in Methods-Time Measurement, R12A means Reach 12 Inches to an object in a fixed location).

4.1.2 Alphabetic and Numeric Codes

Figure 4.2 is an all-numeric code in which numbers are given meaning other than than the value normally associated with the number. The code is typical of codes used in industry. This hierarchical type of code will be treated in considerably more detail later in the chapter.

Figure 4.3 shows a coding system using alphabetic and numeric characters. The alpha characters are mnemonic: that is, the alpha codes represent the first letter of the word. It is a system of codes for planning, measuring, and controlling clerical work, called Master Clerical Data (MCD). Although there may be examples of all-alphabetic codes, none come to mind to use as an example.

MASTER CLERICAL DATA
INDEX
SECTION E

				TMU
E	EYE ELEMENTS			TMU
	D	DECIDE		11
	M	MOVE (Per Inch)		1
	R	READ		
		D	Digit or Letter	
			01 Silent (3 digits or letters)	7
			02 Aloud (per digit or letter)	11
		W	Words (average prose)	
			01 Silent	5
			02 Aloud	11
	S	SCAN		

ERD-01 = 7

Figure 4.3 Alpha (mnemonic) and numeric codes representing a system of predetermined motion times for clerical operations. (Courtesy of Serge. A. Birn, Co.)

4.2 NONCODES

It is important for you to disabuse yourself now of any false notions you may have about codes. The words *code* or *coding* are very often misused or misapplied. To many data processors, any set of characters representing a bit of information is a code. This is not always true. The following sections present examples of noncodes.

4.2.1 Part Numbers (without Interpretive Significance)

Often, these tag numbers are referred to as codes because they are arbitrarily assigned to represent something (i.e., one tag number for one thing). One firm with many divisions worldwide assigned a seven-digit all-numeric symbol to each unique item in their system, or so they thought. There was significant duplication, as investigation later revealed. In this kind of numbering scheme, for example, 5320704 was assigned to a hermetic rotor assembly, while 5320703 was assigned to a screw. The numbers were assigned sequentially in random order (for the most part) of things as created. The only significance was of practically no use at all. The numbers assigned within a block allocated to division "D," identified the division where the parts originated. In no way does the number describe any attribute of the part it represents. Therefore, such part numbers are not codes because they cannot be deciphered to reveal the classification parametric attributes.

4.2.2 Where Used

Frequently, part identifier systems relate the part to the model on which it is used (where used information). It is important to know on what assemblies and models a part is used, but—ah, there's the rub. If a part is used on one model and bears that model's identifier as part of its part number, what happens when the same item is used on another model?

We have found most generally that firms choose one of two solution—both bad. One is to give the same part two or more different numbers, one for each model on which an item is used. The other is to keep the first use number, if it can be found. Even so, this is contrary to the intent of the system as developed and requires some form of cross reference.

In neither case does the part number identify any of the attributes of the part itself. "Where-used" is essential information but a nonpermanent characteristic. Further, the bill-of-materials processing system offers the basis for consolidating the requirements in a horizontal search of the file of bills. The data base is searched for all models and/or assemblies in which a given part is used. In other words, the system—not the part number—should produce the desired where-used information. Where-used is not a code in this context.

4.2.3 Drawing-Paper Size

Another frequently encountered system of identification is that which utilized the size of the paper upon which the part is drawn. When an engineering change is to be made, it is necessary to know paper size if the part drawings are filed by size (all A size together in a file, all B size together, and so on). But part of the identifier? No way; and especially so with computer-generated designs!

4.2.4 Configuration Level

The same can be said for parts identifiers that show configuration level as an integral part of the identifier. The information is useful and may be essential, in some cases of unitary configuration control, but even this is debatable. If a part, after change, is completely interchangeable with the prior changed part configuration (i.e., old works in new, new works in old), the number remains the same. If, on the other hand, old will interchange with new but new will not interchange with old, or vice versa, then the part number identifier must change.

For example, the last two code positions in one firm's numbering system are reserved for change level, irrespective of the nature of the change. One change we found affected only the position where the part number was stamped. Another change relieved the tolerance on one noncritical dimension. None of these changes affects interchangeability, yet there were two part number changes (i.e., the last two digits changed).

There are similar change-configuration problems with assemblies and models. Configuration changes affect model designations just as they do parts, materials, and components. There are many problems to be overcome that are peculiar to product coding—just one of which is the effect of engineering changes on componentry and/or assemblies that make up any one end product.

Assemblies are subject to changes imposed in some cases by a supplier of commercial parts. One example concerns stepping motors used by a maker of computer printer peripherals. The vendor's catalog items were barely meeting the specifications laid down by the firm making printer. These specs were taken from the maker's catalog options. The supplier increased the cost to pay for preselection, in much the same manner that bearing makers preselect standard bearings for ultra-quiet operation for some electric motor makers. The customer, the printer maker, refused to accept the increase and found another vendor whose product, although meeting the requirements, was made using metric measure. This required changes in mountings, couplings, and connections, which originally had been designed to accept U.S. Customary measure parts. The end assembly performance was exactly the same as before, even though the parts were changed. The assembly number was changed so that the service support for the model could be maintained. In that particular identifier system, even though the assemblies were interchangeable, it was thought necessary to change the assembly number. We recognize the problem, but there is a better solution using either the serial number and/or the date of manufacture. The part number should have remained the same.

We are trying to point out that part numbers that reflect the mark level or change level are not codes any more than drawing sizes or random numbers—meaningless tag numbers—are codes, as we have defined them.

What, then, is a code in the context in which we are using it, in a practical sense?

4.3 INTEGRATED CODES

We will not belabor the codes described in Section 4.1, as they are:

Already established and in use for the stated objectives not necessarily related to data coding per se.

They relate to various populations of things, some quite apart from the problem of areas we are attempting to organize and resolve.

There are basically two coding methods used in industry to code data. One is the *integrated code,* based upon hierarchical classifications of data. An integrated code is bound by the scope of the classification and the population of things it must embrace. This type of code requires the most skill in development, and considerable practice opportunity for it to be done efficiently.

In 1954, we coined a word to designate the integrated, hierarchically oriented code from the less complex, simpler coding forms. Codes that are integrated with the heirarchical classification we call *monocodes.* The second coding form includes codes that are not integrated within the hierarchy of the classification. They may, however, relate to a fixed field type of classification. We call these *polycodes.* These terms are copyrighted by E. G. Brisch and Partners Ltd. (now Brisch, Birn and Partners), but the profession generally has adopted these names as generic expressions. The Brisch firm has not pursued the protection afforded by the copyrights.

We define a *monocode* as follows:

> An integrated code of fewest characters which distributes a population of data in a reasonably balanced, logical, and systematical order, where each code character qualifies the succeeding code.

In the example in Figure 4.4, 3285-007 and 4285-007, the digit 2 that follows the digit 3 has an entirely different meaning than the digit 4 in the first code position. The digit 8 is qualified, therefore, by the preceding digit 2 in the second code position, and so on.

The family code corresponds to the classification family—that is, a numeric designation of the meaningful, permanent characteristics of the data being coded. In Figure 4.4 it is 3285 in the one case and 4285 in the second. Expressed in format language, we can establish a hierarchical code order, as Figure 4.5 shows.

The family may consist of three, four, five, or even six digits. The size of the family code is intimately related to and dictated by the

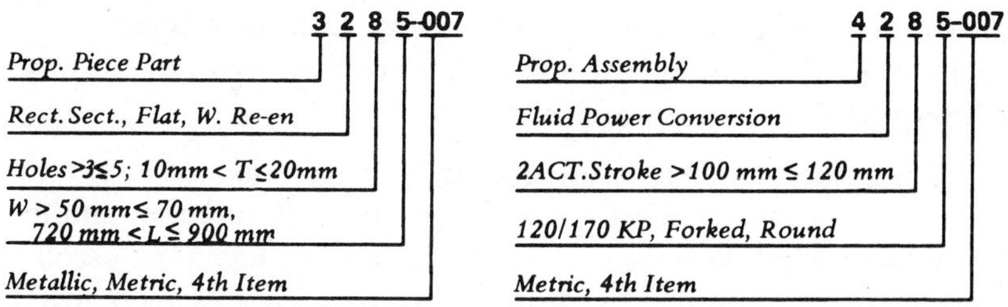

Figure 4.4 Explaining the integrated code.

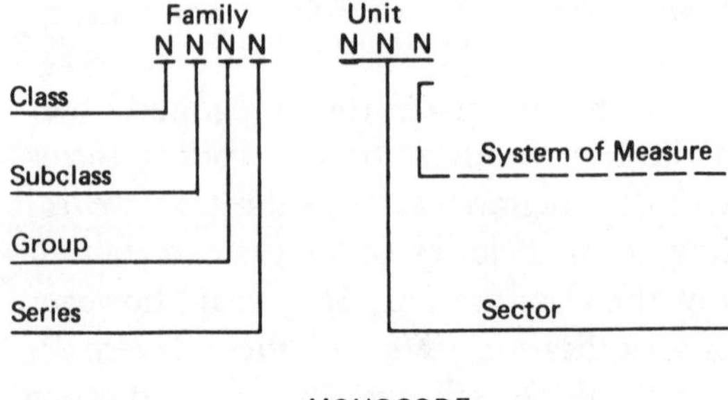

Figure 4.5 Classification-based integrated hierarchical codes.

number of items embraced by the classification, now and in the longer 20–25 year term.

The individual members of the family are arrayed by their essential differences in the sector. Additional coding meaning can be imparted to the limits that make up the family (the portion of the code format referred to as the sector) in a number of ways.

1. By use of the even-odd designation shown in Figure 4.5, where it is used to designate system of measure.
2. By a sector block allocation for major configurations within a family, called *sectorizing* (e.g., 001/199 for plain finish, 201/299 for brass finish, or 001/499 for U.S. divisions and 501/999 for foreign divisions).

A reasonably balanced distribution is desirable because:

There must be enough data within a family to provide a reasonable choice; *but*

Not so many as to increase unnecessarily the search and retrieval time and/or require another step in the retrieval process to break down the items further, into yet another listing or charting of data which will require even more search time.

Computer software has been developed to catalog certain data when, for purposes of simplification and standardization, large families are tolerable—in fact, desirable. Once simplification has been accomplished, the numbers of items are usually 30–60% fewer. This reduced variety thereby keeps the search and selection time within the range

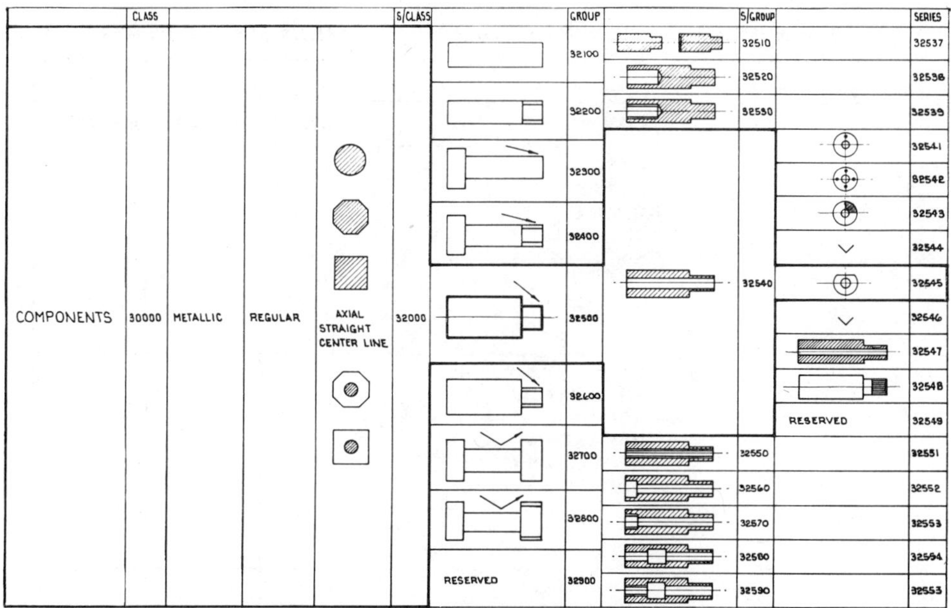

Figure 4.6 Integrated monocode.

30–150 sec. This subject (Classification and Coding and the Computer) is treated in some detail in Chapter 8.

We often multiplex the coding forms (monocodes and polycodes) to produce what we refer to as multicodes, in accordance with certain established principles. Figure 4.6 shows the basic means of conveying information using polycodes: that is, as a series of independent data items relating to an established file of records, or a subfile of records within a file. It can be composed of alpha, numeric, or alphameric code characters. Multiplexing the two coding forms simply means trailer coding the polycode data behind the record file address represented by the monocode. Figure 4.7 shows how the coding schemes interface—again, following certain prescribed principles.

4.4 PRINCIPLES OF CODING

Just as there are principles for industrial classification (and all classification work), so there are principles for the code developer to observe. E. G. Brisch and Partners Ltd., working in consultation with the

CODE NUMBER	DESCRIPTION	ANALYSIS OF FEATURES		
3 2 2 9 3 3 -101		a	d	e
3	COMPONENT	MAT'L. SHAPE	MAX. DIA.	O/A LENGTH
2	METALLIC REGULAR FORM STRAIGHT C/L. HEXAGONAL OR ROUND SINGLE OR MULTI-O.D.			
2	WITHOUT CENTER HOLE WITH THREAD			
9	MAX. SECTION BETWEEN ENDS OTHER THAN PLAIN			
3	ROUND TWO OR MORE THREADED PORTIONS	a_1 ROUND BAR		
3	O.D < 2" LENGTH > 24"		d_1 1.375 MAX.	e_1 30.625
-101	SECTOR NUMBER (Seq.)			

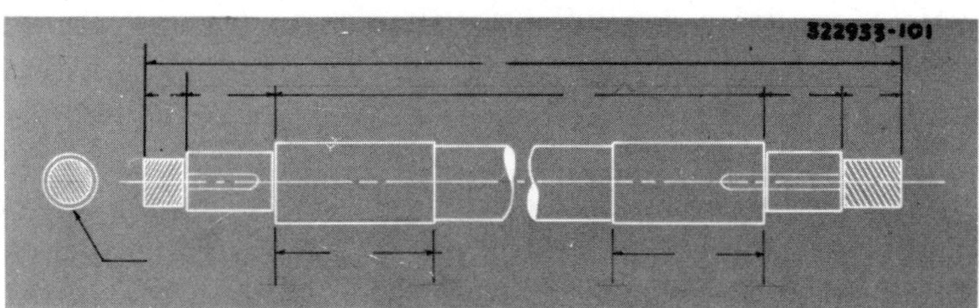

Figure 4.7 Interface of monocodes and polycodes.

BY MEANS OF DESCRIPTORS

g	h	i	k	α	β	γ
HOLES	THREAD	MAT'L CODE	KEYWAYS	TURNING DIA.(in.)	TURNING LENGTH(in.)	DRILLING DIA.(in.)
		i_1 SAE 1035		α_1 0.500	β_1 0.687	
		H.R. 11361-269		α_2 0.8745	β_2 1.625	
				α_3 1.375	β_3 2.500	
		OR		α_4 1.3437	β_4 20.000	
		i_2 SAE 1045		α_5 1.375	β_5 3.000	
		H.R. 11364-269		α_6 0.8760	β_6 1.625	
				α_7 0.750	β_7 1.187	
	h' U.N.C.					
g_1 ONE			k_1 2			γ_1 $\frac{3}{16}$"
g_2 RADIAL			k_2 0.1875" WIDE			
			k_3 1.375" LONG			
	h_1 $\frac{1}{2}$-13 UNC					
	h_2 $\frac{3}{4}$-10 UNC					

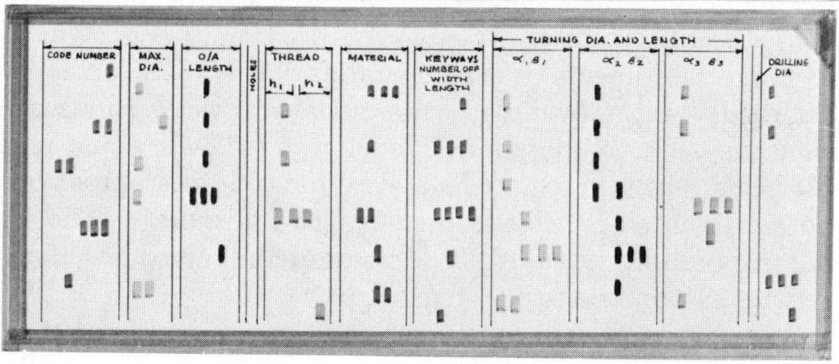

TURNING DIA. AND LENGTH

CODE NUMBER | MAX. DIA. | O/A LENGTH | HOLES | THREAD h_1 h_2 | MATERIAL | KEYWAYS NUMBER OFF WIDTH LENGTH | α_1 β_1 | α_2 β_2 | α_3 β_3 | DRILLING DIA

leading authorities in the fields of codes, coding, and short-interval memory-retention problems, laid down these principles for its own professionals to use. Both the principles and their corollaries are the only guidances that we know of that have been recorded. Much of our experience is contained in these principles, as well as those of Dr. R. Conrad, of the United Kingdom. They are as follows:

PRINCIPLES OF INDUSTRIAL CODING

1. No code should exceed five charcters without a break in the string. COROLLARY: The shorter the code, the fewer errors.
2. Identity codes should of fixed length and pattern. COROLLARY: Varying-length codes within a given class of materials proliferate error rates and require justification in handling (right or left) to the longest code in use.
3. All-numeric codes produce fewest errors. COROLLARY: A fixed-length, all-numeric code of five or fewer digits is best.
4. Alphanumeric combination codes are acceptable if the alpha field is fixed and used to break a string of numbers. COROLLARY: Alpha and numeric codes intermixed in the same code position cause excessive transaction errors.

We have alluded to the use of alpha characters with certain restrictions imposed. That is, we have used alpha characters in a code position, but only under conditions where the alpha character defines a category of data in a useful manner.

In product coding projects, for example, it is often necessary to show the model year. Obviously, if the model year were expressed with two characters—say, 78 to indicate model year 1978—it would require two code fields. The firm required a 20-year life cycle on its product support, so an alpha character was used to good advantage. The alpha character was used in the fourth code position and served as a break to an otherwise all-numeric string of six digits.

4.5 APPLYING MONOCODES

The method of applying the monocode procedure can best be demonstrated by the example of the attribute blocks. If you will recall, the classification options were several: all items of the same color together, all items of the same shape together, and so on.

Let us classify the population into families of items that are the same size, shape, thickness, and color. We can code this classification in several ways:

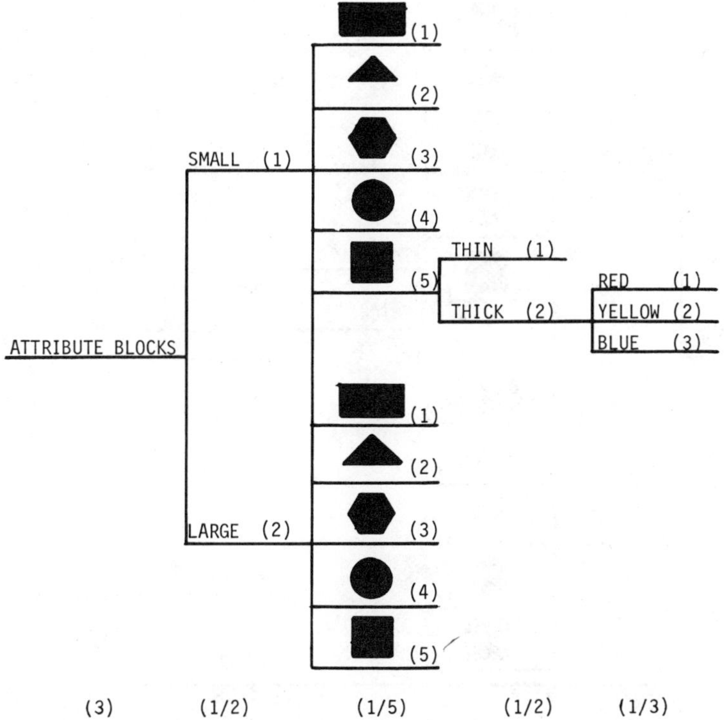

Figure 4.8 Polycoding the classification: one parameter, one code.

1. To reflect one code position for every level of the classification (i.e., polycoding each parameter). Figure 4.8 shows how this is done.
2. To compress the data into an integrated form; that is, two levels are compressed into one code positon. In Figure 4.9 you can see how one code position encompasses two parameters without compromising retrievability or the unique identity of the individual items (i.e., monocoding).

Summarizing: A small, square, thick, yellow block can be polycoded 3142-1 (Figure 4.8) or monocoded 34-5 (Figure 4.9), with each uniquely identified. The shorter code is preferable because it requires two fewer digits to accomplish the identical result (i.e., like coded things are brought together by virtue of their similarities, and uniquely identified by their essential differences).

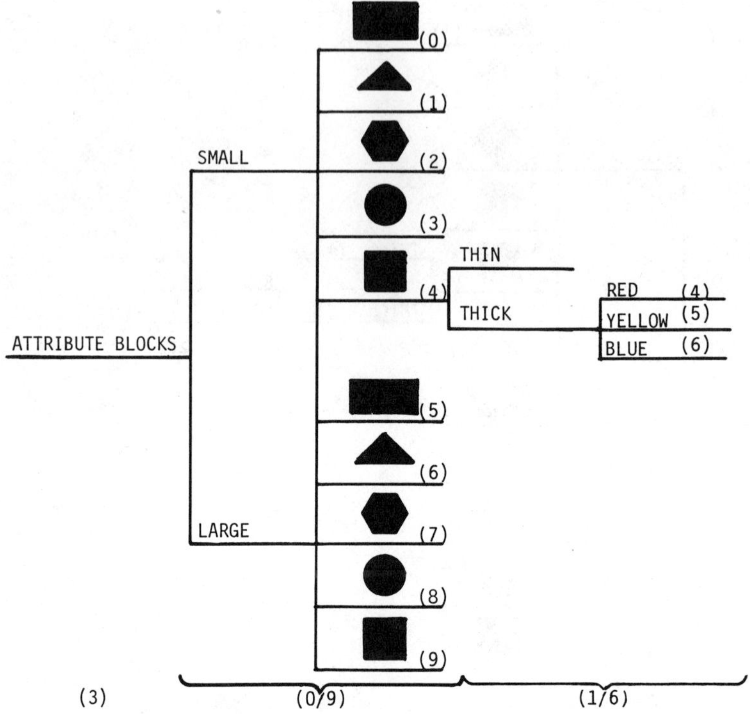

Figure 4.9 Monocoding the classification.

If three digits will identify uniquely each item, it makes little sense to use five. If you recall, the monocode captures considerable data in fewest digits. You will also recall that an all-numeric, fixed-length code produces fewest errors; and the less data there are to store.

Now, let us consider a more involved problem, applying the same principles to a classification of a more complex population of items. In Chapter 3 we used an illustration of three round parts. These parts were extracted from an active population of 24,000 + parts of many diverse forms, functions, and/or features. Figure 4.10, an illustration of six parts, simulates the distribution of parts in very general terms. How would a classification-based code be developed to reflect this distribution?

1. Capture all 24,000 parts and copy them on a standardized paper size for ease of processing.

2. Develop a classification-based analysis code after a preliminary examination of the population of the items. This is generally done from the findings in the preliminary survey from the 2, 3, or 4% sample that was used to determine the design prevention rate (see Chapter 2).
3. Sort the 24,000 (or however many are considered active in your file of designs) to the three digits that comprise the categories of Figure 4.11.
4. Quantify the distribution and determine the areas of the classification (and code) that must be expanded (i.e., must reveal more definitive parameters in the classification) or where to contract the classification.
5. Synthesize a tailor-made classification tree that reflects the population distribution.
6. (a) Distribute the items to the new tree and then to cybernet sheets which display the classification tree and code for manual use. (b) Enter tree into computer for electronic handling.
7. Audit the family members to ensure that (a) they are correctly classified; (b) there is a reasonably balanced distribution of the total population.
8. Code by marking the family numbers on the drawing copies. In Figure 4.12(a), the drawing of the part is identified only to family level, whereas the designed part shown in Figure 4.12(b) is uniquely identified.
9. Catalog the items in selected families for simplification studies.

There are other steps relating to processing the data, storing media, and the introduction of variety controls, but these subjects are treated in later chapters.

Distribute the population of items represented in Figure 4.10, items A through F. To do this, photocopy the items on the pages, separate them, and assign a three-digit sorting code to each of the six items, using the sorting code shown in Figure 4.12. You should have the following codes:

A = C36
B = C61
C = C21
D = C62
E = C25
F = C41

Quantify the distribution. You are now ready to synthesize the

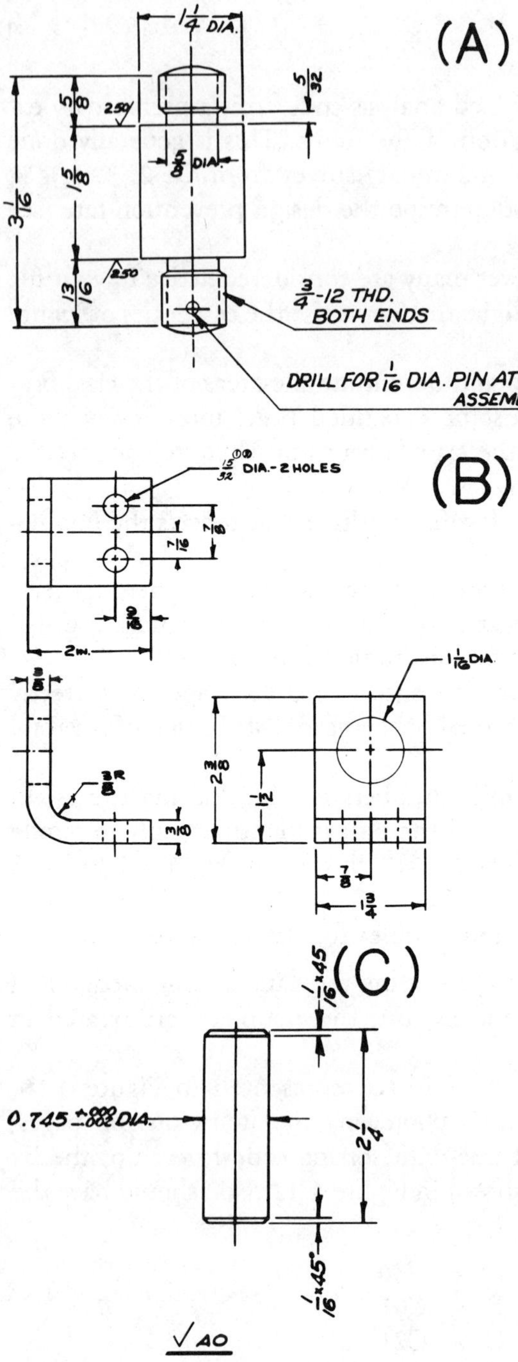

Figure 4.10 Six piece part drawings.

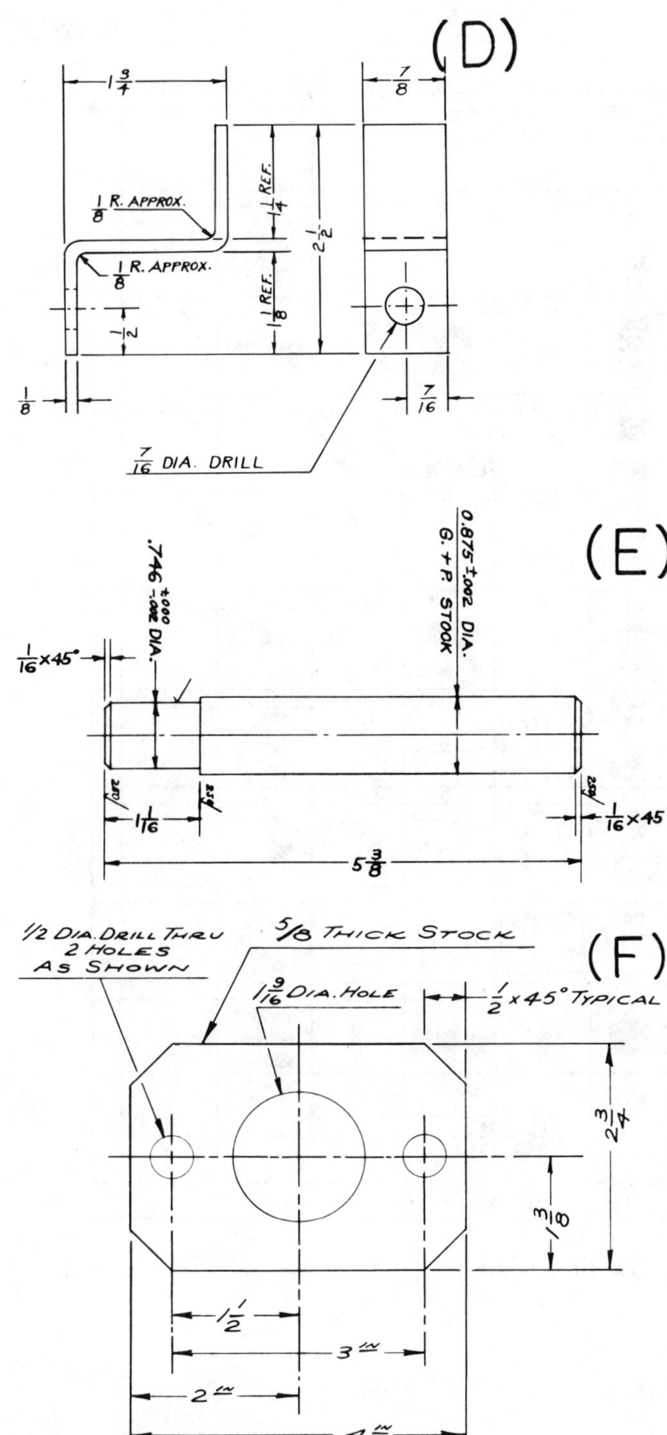
(D)
1 3/4
7/8
1/8 R. APPROX.
1 1/4 REF.
2 1/2
1/8 R. APPROX.
1 1/8 REF.
1/2
1/8
7/16
7/16 DIA. DRILL
(E)
.746 +.000 -.002 DIA.
0.875 ±.002 DIA.
G. + P. STOCK
1/16 × 45°
1 1/16
1/16 × 45°
5 3/8
(F)
1/2 DIA. DRILL THRU
2 HOLES
AS SHOWN
5/8 THICK STOCK
1 9/16 DIA. HOLE
1/2 × 45° TYPICAL
2 3/4
1 3/8
1 1/2
3 IN
2 IN
4 IN

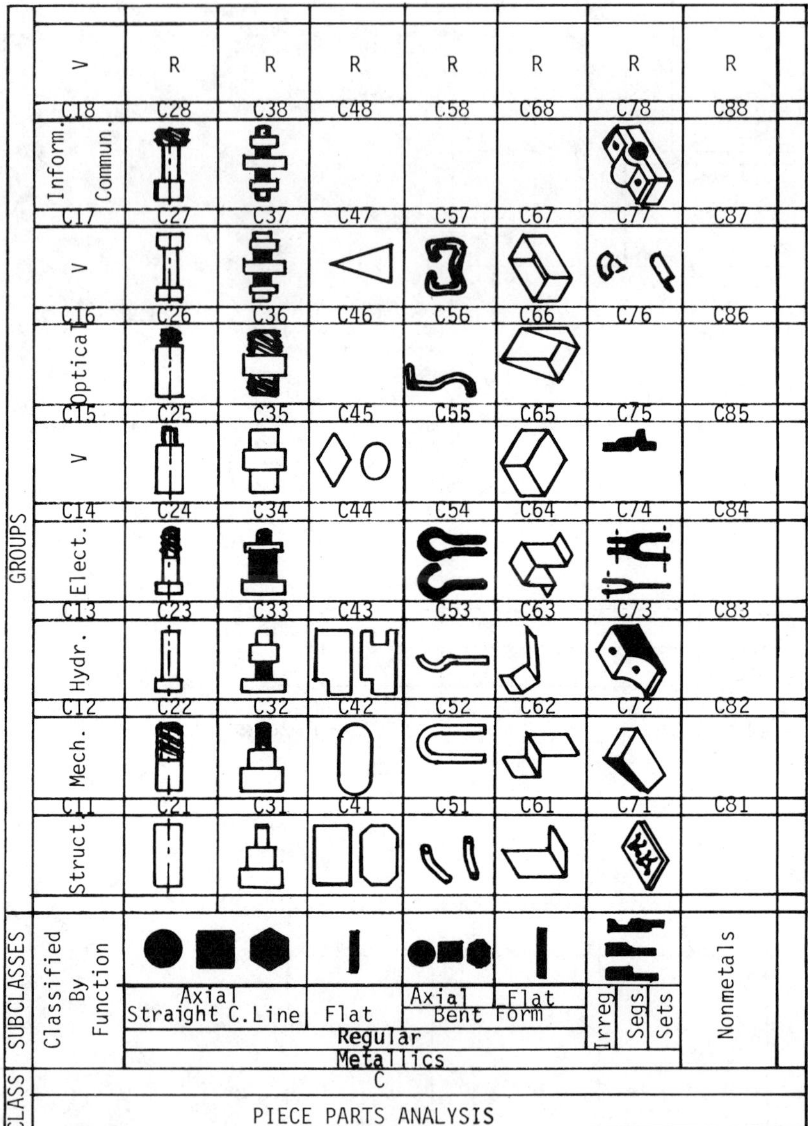

Figure 4.11 Sorting code for analysis.

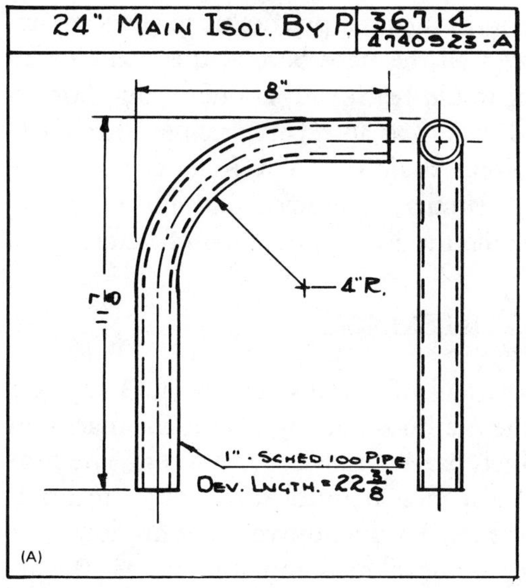

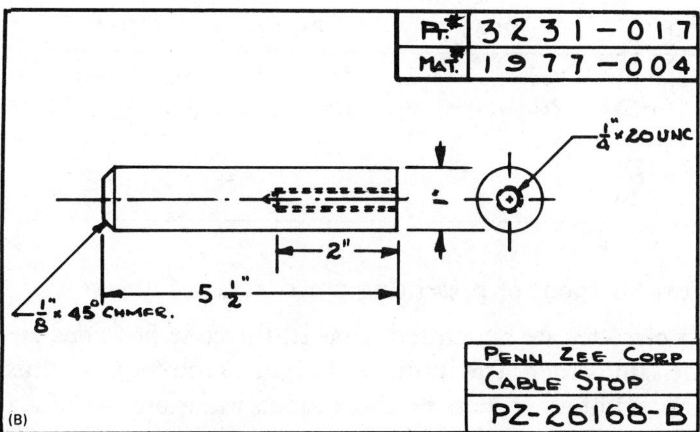

Figure 4.12 Part coding: (A) coded to family level; (B) uniquely identified by family and sector number.

final classification and monocode form. It is not practicable to create an exercise that would take you through all the steps beyond this point, or to simulate a classroom exercise that would teach you, at one sitting, how to develop logic patterns that would make you an expert classifier. It requires at least 8 months of intensive training to do this. This is not the object of this book. But what is shown is a simulation exercise that illustrates how classifications are tailor-made to reflect a given population of items.

4.6 POLYCODING AND MULTIPLEXING

Polycodes are developed for a variety of reasons. Figure 4.13 shows a polycode system for recording the methods making a part, the machineability of the materials from which parts are made, and so on. The firm that developed this set of polycodes manufactures high technology—usually airborne—gas turbines. Production volumes are low.

To illustrate how polycodes are developed and applied, use the attribute block set once more. The polycode arrangement can be in any order, independent of any other parameter.

P1(1/5 or A/E)			P2(1/3)			P3(1/2)			P4(1/2)		
Round	=	1 (or A)	Red	=	1	Thin	=	1	Small	=	1
Square	=	2 (or B)	Yellow	=	2	Thick	=	2	Large	=	2
Rectangle	=	3 (or C)	Blue	=	3						
Triangle	=	4 (or D)									
Hexangle	=	5 (or E)									

There are two methods of presenting polycodes, as follows:

1. *Fixed-field* polycodes are structured; that is, the code positions are always in the same order. The number of digits in the code are thus equal to the number of parameters and/or variations within a parameter.
2. *Variable-field* polycodes require a parameter identifier. These are open-ended; that is, the number of polycodes may vary.

In the preceding example, P1 represented shape, P2, the color, and so on. Because the order was known and the fields fixed, code 1 1 1 1 represents a round, red, thin, small block. In a variable-field form, where the parametric code order is not necessarily maintained (common practice in a data-base system), it is necessary to identify the record bits further. The same polycode, 1 1 1 1, in a fixed-position structured code in a file would read 11 21 31 41. Then, regardless of its randomly ordered position in the file, each record bit is retrievable.

A more realistic version of polycodes applied to the solution of a specific management problem is illustrated by the following example. The object of the project was to improve and speed customer and sales service at order entry. A polycode was multiplexed to a product monocode. The firm's products were (and are) gear boxes. Upon customer or salesperson inquiry for a quotation, the product type was monocoded. The variety within any one family could be quite large because of custom requirements. The product monocode was kept to five characters—a radical departure from the usual 20-character industry code. A manual search system was developed to cull the 300 or more family members so that response was kept within a 3-min long-distance basic call cost unit. The system produced cost, lead time, and delivery information. The product code was applied in 30 sec. A further screening within the family required less than 1 min to:

Determine whether or not an exact duplicate had been engineered and made; or

Retrieve the nearest item similar to the one requested.

As can be appreciated, this system now uses the computer and a terminal for magnetic storage, access, and retrieval. The manual system was used because there was no computer in-house at the time of its development.

Another use of polycodes is to trail data behind a file address for nonpermanent characteristics that are either dynamic and/or not needed for continued data handling.

Some years ago, one firm asked to have all purchased nonproprietary items coded to reflect:

The presence of copper in copper-based alloys in the product.

The approximate weight of the copper or copper-based alloys in the items.

The reason was quite valid. At the time, copper prices were fluctuating wildly. It was not unusual to have $0.05-per-pound change on copper wire bar prices on the London Metal Exchange. The purchasing director was getting calls from suppliers of, say, electric motors advising of a price escalation "due to a rise in copper prices." The purchasing manager wanted:

Category	0	1	2	3	4	5	6	7	8	9
MATERIAL – CHEMISTRY	NONFERROUS METALS			FERROUS METALS			SUPER ALLOYS			O/T A0/A8 INCLUDING COMPOSITES A9
				NON CORROSION RESISTANT			PRINCIPLE ELEMENTS			
	ALUMINUM & ITS ALLOYS A0	MAGNESIUM & ITS ALLOYS A1	O/T A0/A1 TO INCLUDE TITANIUM A2	4300 SERIES A3	CARBON STEEL ALLOYS. 6260, A4 6265 ETC	STAINLESS A5	NICKLE. INCO, WASP, HAST. A6 ETC	COBALT INCL. STELLITE A7	O/T A6/A7 A8	
MATERIAL – FORM & SIZE	PREDETERMINED SIZE			ROUND BAR STOCK INCLUDING TUBING – DIAMETER						O/T B0/B8 INCLUDING BAR STOCK SHAPES ■●▲ B9
	CASTING B0	FORGING B1	EXTRUSION B2	≤ 1" B3	>1" ≤ 2" B4	>2" ≤ 4" B5	>4" ≤ 6" B6	>6" ≤ 10" B7	>10" B8	
	PART LENGTH									
	≤ 1" C0	>1" ≤ 2" C1	>2" ≤ 4" C2	>4" ≤ 6" C3	>6" ≤ 8" C4	>8" ≤ 10" C5	>10" ≤ 14" C6	>14" ≤ 18" C7	>18" ≤ 28" C8	> 28" C9
MATERIAL – TREATMENT – METALURGICAL	O/T HARDEN			HARDEN CASE						
				CARBURIZING						
				ALL OVER		SELECTIVE				
	ANNEAL ONLY D0	STRESS RELIEVE ONLY D1	NORMALIZE ONLY D2	TYPE I D3	TYPE II D4	TYPE I D5	TYPE II D6	O/T CARB TO INCLUDE NITRIDE D7	INDUCTION D8	FLAME D9
	THROUGH HARDEN									
		HEAT, QUENCH, DRAW – MAX HARDNESS								
	UNSPECIFIED E0	≤ 32 R_c E1	> 32 R_c E2	MARTEMPER 5045 E3	PRECIPITATION OR AGE 5038, 5071 E4	STABILIZE WASPALOY E5	D3/D4/D5/D6 + E0/E1/E2/E3/E4/E5 E6	D8 + E0/E1/E2/E3/E4/E5 E7	E8	NONE OR AS RECEIVED E9
MATERIAL – TREATMENT – SURFACE		CASE/PLATE/INFUSE SURFACE					WORK SURFACE			
			ELECTROPLATING							
	NONE OR AS RECEIVED F0	ANODIZE INCLUDING PARKINIZING F1	COPPER F2	CHROME INCLUDING CU/NI F3	LEAD/GOLD F4	PEEN OR BLAST F5	POLISH F6	HIP F7	AF5411 F8	O/T F0/F8 F9
PROCESS – FORM – SURFACES – EXTERNAL ONLY – OD'S – TURN ONLY		FROM ONE OR BOTH ENDS								
		STRAIGHT TURNS – NO. OF SURFACES			TAPERED TURNS – NO. OF SURFACES					
	NONE OR AS RECEIVED G0	≤ 2 G1	>2 ≤ 4 G2	> 4 G3	≤ 2 G4	> 2 G5	G1/G2 + G4 G6	G2/G3 + G4 G7	G2/G3 + G5 G8	O/T G0/G8 G9
PROCESS – FORM – SURFACES – EXTERNAL ONLY – OD'S – O/T TURN ONLY		GRIND					COMBINATIONS			
			FINISH – NO. OF SURFACES							
	NONE REQUIRED H0	FORM LABYRINTH SEALS H1	≤ 2 H2	> 2 H3	CRUSH FORM O/T LABYRINTH H4	H1 + H2/H3 H5	H1 + H4 H6	H2/H3 + H4 H7	H1/H2 + H3/H4 H8	O/T H0/H8 H9
PROCESS – FORM – SURFACES – EXTERNAL ONLY – ENDS ONLY	TURN						COMBINATIONS			
	FACE INCL. CHAMFER RADIUS J0	FACE + TREPAN J1	GRIND SURFACE(S) J2	MILL J3	CURVIC ONE END J4	CURVIC BOTH ENDS J5	J0/J1 + J2 J6	J0/J1 + J3 J7	J6 + J7 J8	O/T J0/J8 J9
PROCESS – FORM – SURFACES – INTERNAL ONLY – WORK TURNS		TURNED/BORED SURFACES TO BE MACHINED					COMBINATIONS			
		NOT GUN DRILLED OR GROUND – NO. OF SURFACES								
	NONE OR USE AS RECEIVED K0	≤ 2 K1	>2 ≤ 4 K2	> 4 K3	GUN DRILLED K4	GROUND K5	K1/K2/K3 + K4 K6	K1/K2/K3 + K5 K7	K5 + K6 K8	O/T K0/K8 K9

PROCESS FEATURES		0	1	2	3	4	5	6	7	8	9
PLAIN HOLES	TYPE		TRAVERSE ONLY	TRAVERSE ONLY	LONGITUDINAL ONLY	LONGITUDINAL ONLY	LONGITUDINAL ONLY	LONGITUDINAL ONLY	COMBINED	COMBINED	
			PERPENDICULAR TO ℄		PARALLEL TO CENTER LINE	PARALLEL TO CENTER LINE	PARALLEL TO CENTER LINE				
			SPACING		ON BOLT CIRCLES	ON BOLT CIRCLES	ON BOLT CIRCLES				
					SPACING ONE BOLT CIRCLE	SPACING ONE BOLT CIRCLE					
		EQUAL INCL. ONE HOLE L0	NOT EQUAL L1	SKEWED L2	EQUAL INCL. ONE HOLE L3	NOT EQUAL L4	TWO OR MORE BOLT CIRCLES L5	ASKEW L6	WITH EQUAL PATTERNS L7	WITHOUT EQUAL PATTERNS L8	O/T L0 / L9 INCL. NONE L9
	PROCESS		PROCESS FOR CREATING HOLES	PROCESS FOR CREATING HOLES	PROCESS FOR CREATING HOLES	PROCESS FOR CREATING HOLES	PROCESS FOR CREATING HOLES	PROCESS FOR CREATING HOLES	PROCESS FOR CREATING HOLES	PROCESS FOR CREATING HOLES	
		NONE OR AS RECEIVED M0	DRILL M1	DRILL + REAM/TAP M2	DRILL/REAM + HONE M3	DRILL + GRIND M4	PUNCH M5	EDM OR ECM M6	BROACH M7	LASER M8	O/T M0/ M9 INCL. COMBI. M9
THREADED SURFACES			O.D. ONLY	O.D. ONLY	O.D. ONLY	O.D. ONLY	I.D. ONLY	I.D. ONLY	COMBINED THREADED SURFACES	COMBINED THREADED SURFACES	
		NONE OR AS RECEIVED N0	CHASE N1	SINGLE POINT TURN N2	GRIND N3	ROLL N4	TAP N5	SINGLE POINT BORE N6	≤ 4 N7	> 4 N8	O/T N0/N8 N9
O/T PLAIN GEARS	EXTERNAL		SINGLE INVOLUTE GEAR TOOTH FORMS	SINGLE INVOLUTE GEAR TOOTH FORMS	SINGLE INVOLUTE GEAR TOOTH FORMS	SINGLE INVOLUTE GEAR TOOTH FORMS	SINGLE INVOLUTE GEAR TOOTH FORMS	SINGLE INVOLUTE GEAR TOOTH FORMS	SINGLE INVOLUTE GEAR TOOTH FORMS		
			SPUR GEARS	SPUR GEARS	SPUR GEARS	BEVEL GEARS	BEVEL GEARS	HELICAL	HELICAL		
			PRESSURE ANGLES	PRESSURE ANGLES	PRESSURE ANGLES	PRESSURE ANGLES	PRESSURE ANGLES	PRESSURE ANGLES	PRESSURE ANGLES		
		NONE OR AS RECEIVED P0	< 20° P1	≥ 20° < 30° P2	≥ 30° P3	< 20 P4	≥ 20° P5	< 20° P6	≥ 20° P7	MULTI OR CLUSTER GEARS P8	O/T P0/ P8 TO INCL. PAWLS P9
	INTERNAL		SINGLE INVOLUTE GEARS	SINGLE INVOLUTE GEARS	SINGLE INVOLUTE GEARS						
			SPUR GEARS	SPUR GEARS	SPUR GEARS						
			PRESSURE ANGLES	PRESSURE ANGLES	PRESSURE ANGLES						
		NONE OR AS RECEIVED R0	< 20° R1	≥ 20° < 30° R2	≥ 30° R3	MULTI OR CLUSTER GEARS R4	R5	R6	R7	R8	R9
SERRATIONS SPLINES KEYWAYS					SPLINES	SPLINES	SPLINES	SPLINES	SPLINES	COMBINATIONS	COMBINATIONS
					EXTERNAL	EXTERNAL	EXTERNAL	INTERNAL	INTERNAL		
		NONE OR AS RECEIVED S0	SQUARE HOLE D-HOLE & KEYWAYS S1	SERRATION ONLY S2	HOB S3	SHAPE S4	O/T S3/S4 INCL. PUSH BROACH S5	PULL OR PUNCH BROACH S6	SHAPE S7	S1/S2 + S3/S7 S8	S3/S5 + S6/S7 S9

Figure 4.13 Manufacturing polycodes.

To know if the price-change amount was valid.

The capability to call vendors when the price of copper was reduced in in order to have the price of the product reduced accordingly.

Added, as additional record data, were the following polycodes:

P1 = presence of copper
10 = no copper or copper alloy
11 = electrolytically pure conductor-grade copper only
12 = copper not conductor-grade only
13 = brass only
14 = bronze only
15 = 11 plus 13
16 = 11 plus 14
17 = 12 plus 13
18 = 12 plus 14
19 = 13 plus 15
P2 = weight of copper or copper-based alloys
= in pounds per item
20 = <0.5
21 = ≥0.5 to <1.5
22 = ≥1.5 to <2.5
23 = ≥2.5 to <4.0
24 = ≥4.0 to <6.00
25 = ≥6.00 to <8.00
26 = ≥8.00 to <10.00
27 = ≥10.00 to <13.00
28 = ≥13.00 to <18.00
29 = ≥18.00

Also added were existing vendor codes in a file merging program, keying on the monocode of the item and its existing part number. They then printed all suppliers and the amount of materials purchased per lot (from past history by account) to provide the data needed. In approximately 8 months, after the Chilean political upheaval subsided, such data were no longer needed. The program was retired for the moment; the disk-stored data were kept, however.

It is well to remember that polycodes are easily developed if:

1. The objectives are well defined.
2. The composition of the body of data is known and fully understood.

Most universal systems of codes are polycodes and, because they must cater for the unknown, the numbers of digits in the code are usually excessive. A good example was an industry code proposed by a committee comprised of members of firms of fastener manufacturers. It was a system of polycodes, 27 characters in length, to describe a nut, bolt, or screw, plus three characters for pack. The permutations are in the trillions, with most of them useless because the combinations are improbable. With this code, one could code a No. 10–42 round-head slotted screw 999.99 in. long, made of glass and coated with platinum. Had the system been adopted, it would have been obsolete because the system of measure was U.S. Customary. Wiser heads prevailed and the proposed code dropped. Whoever said that a camel was a horse designed by a committee must have attended a code development committee meeting!

The reference earlier to the code for denoting the presence of copper in purchased items was, in fact, an example of multiplex coding. That is, the polycode used the file address provided by the monocode for effective processing of the data.

It is important to recognize that the form/function classified parts are essentially catering for design retrieval by an enquirer interested in seeing the population of prior designs of parts that can satisfy design requirements. The manufacturing engineer, on the other hand, is interested in the form/function features only to determine the options available to make the part.

Design features, that is, characteristics of the part other than its fundamental exterior and interior configuration, may be common to a variety of differently shaped parts. Some typical design features include the following:

Knurls
Machined flats
Keyways and splines o/t involute splines
Holes o/t center holes on axial parts
Grooves treppanned in a face
Threaded surfaces o/t on centerline
Nonsquare faces

All the parts in subclasses 32 and 33 shown in Figure 4.10 can have design features in common, irrespective of the shape. These features are, of course, of interest to the product engineer but are of equal importance to the manufacturing engineer, who must determine how they are

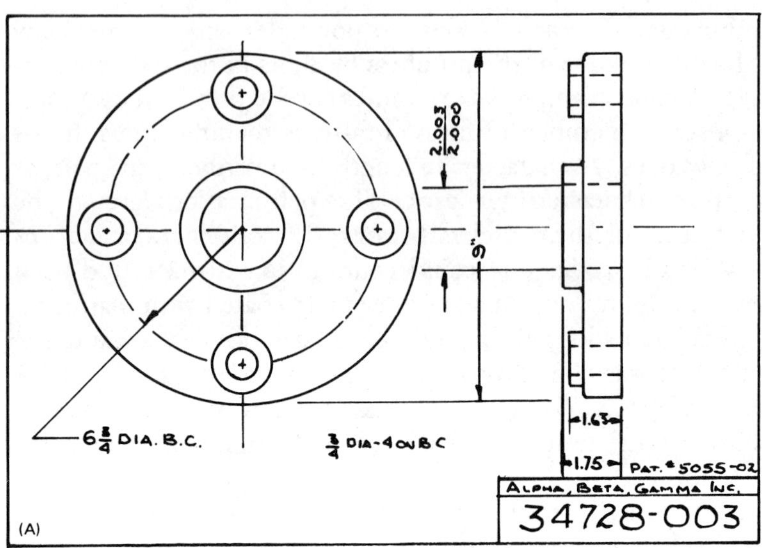

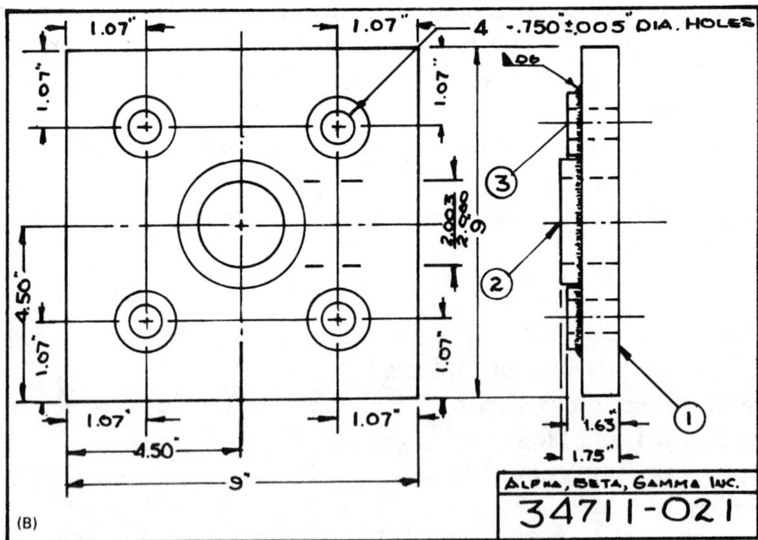

Figure 4.14 (A) Cast part: cover plate. (B) Weldment: box cover. (C) Part machined from steel plate: end plate.

to be made. It is entirely correct to develop a set of polycodes for the use of the manufacturing engineer (or the purchasing department, shipping department, or whomever) that relates the features but emphasizes the processing data [i.e., how the parts will be held, the relationship to the cutting tools (tool rotates versus work rotating) and the like].

Although some of the material in this chapter is covered in more detail in Chapter 7, it is relevant to set the stage in this discussion on polycoding.

Polycoding shape and design features but not how they are created is poor procedure, for the following reasons:

1. Manufacturing methods are dynamic, not static, and are nonpermanent characteristics. Each year when the capital budget is developed, some new technologically advanced piece of equipment may revolutionize the method of manufacturing.
2. The same, or similar, process can be used for parts of widely disparate shapes.

This is true of the parts shown in Figure 4.14. Except for the founding operations (one is a casting and the other two are made from plate (one of which is a weldment), their processes are virtually identical.

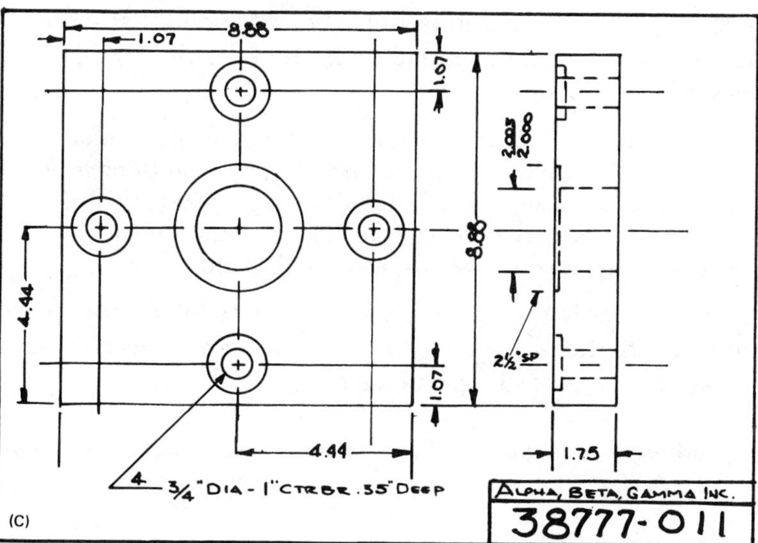

Figure 4.14 (continued)

Therefore, it is well to consider these points when creating sets of polycodes to multiplex with an accompanying monocode. This implies, and correctly so, that the polycodes reflect the processes and processing choices within any one firm's plant(s). Over the years, we have developed a wide variety of these codes. They range from sheet metal products, rolled and extruded products, molded plastics, to, of course, machine shops.

As we said earlier, monocoding requires much skill and practice to develop a logical solution for a given population of data. Polycoding requires considerably less training—several days' practice; a week at most. To multiplex the codes so as to avoid repeating the logic contained in the monocode does require the same logic skills necessary in the development of a monocode. It becomes apparent, therefore, that the use of the multiplex code form can resolve the problems of information retrieval where several users have different needs and viewpoints to satisfy.

The monocode is essentially a vertical-search device, whereas a polycode combined with it enables a horizontal search. This combination is the best method of satisfying *all* viewpoints without compromising either.

John Burbidge,* in summing up the results of the proceedings of an international seminar on group technology, concluded that most experts presenting papers at that conference favored multiplex codes. He said, in his summary of the conference:

> The majority appeared to prefer monocodes for permanent information such as that concerning shape and function, and to prefer polycodes for less permanent information covering a wider range of subjects, such as production method, tooling, lubricants used, annual requirements, used on records, costs, and so on.

Again, while admittedly discussing coding methods in the context of group technology methods, most informed coding experts seem to support this viewpoint on all coding problems.

*J. L. Burbidge, *Proceedings of the International Seminar on Group Technology*, Part A, Turin Center on International Labor Organisation, Turin, Italy, 1973.

5

Simplification and Standardization

5.0 SIMPLIFICATION AND STANDARDIZATION DEFINED

The National Bureau of Standards published a document in 1971 entitled, *An Index of U.S. Voluntary Engineering Standards*. It contained a list of 19,000, or so, recommended practices, testing methods, specifications, and standards, published by American organizations involved in standards work. More than 350 trade associations, technical organizations, and professional societies made up the list of contributors.

Industrial standardization is defined by the American National Standards Institute (ANSI) as follows:

> *Industrial standardization* is the orderly and systematic formulation, adoption, application, and revision of a generally accepted statement of the requirements to be met for the attainment of a recurrent objective.

More directly translated into a firm's needs, ANSI defines company standardization as:

> *Standardization* is a tool for securing optimum utilization of resources and maximum efficiency of operations through formal establishment of the most suitable predetermined solutions to reoccurring problems and needs.

Maxima ex minima! The objectives are virtually the same as the objectives of classification and coding, as well they should.

A company standard is intended to be utilized by all organizational entities within a company that have an effect upon operations: from the concept of design of a good or service, to the delivery of that good or service.

Simplification is an imperative step in the creation of a standard, or a standards program. Simplification means the act of reducing complexity to make easier and may be defined as follows:

> *Simplification* is a management tool consisting of positive actions to reduce the variety of any data to its least common number, commensurate with the demands of the organization.

Simplification and standardization made possible the Industrial Revolution. In 1831, Eli Whitney was awarded a contract to supply firearms to the U.S. government. His firm was to produce 10,000 of the same model of gun. This was a large order for one manufacturer in those days, especially for firearms to be used in the field.

To supply this number of weapons, interchangeable parts became an imperative, not only to make the product efficiently, but to maintain the weapons by field armorers. To appreciate fully the extent of the problem, gunsmithing at that time was an art. Parts were fitted by handwork to mate one another in assembly. It was cut, try, and fit, to match the mating parts. This made each gun a customized, unique item, virtually impossible to repair except by an equally gifted gunsmith.

The success of Whitney and his contemporary, Samuel Colt, at achieving interchangeable standardized parts designs launched the concept of mass production; and the Industrial Revolution in the United States (1831–1845) had begun.

5.1 APPROACHES TO SIMPLIFICATION

To simplify a body of existing data, whether they are gun parts or parts for making computers, requires first that the data be made visible. Only when the variety of data can be seen, arranged in some systematical order, and quantified can anything concrete be done about reducing their variety—if it is to be done in an efficient manner and without harmful effect to the users of the data.

To accomplish simplification may be as straightforward as gathering together in the stockroom one of everything of similar genre in size or some other logical parametric order array. For firms with only a few hundred or so items to manage, this may be a perfectly feasible method. Handling nuts, bolts, rivets, screws, bearings, and seals is easily managed in this fashion. But gathering the components for the manufacture of diesel locomotives presents a somewhat difficult handling problem. For most firms, therefore, except for the easily handled small items, this method of gathering data for simplification study is not only physically cumbersome, but costly and of questionable value in achieving the desired end objective.

A case in point is a firm which, in its several divisions, makes a complementary line of automobiles. To compare one with another (for value analysis and engineering cost comparison purposes and simplification), they physically mounted comparable components and assemblies in vertical alignment. Possible candidates for simplification and/or standardization were visible for comparison and study This was done at their technical center and after the fact—after the cars had been designed, tested, processed, tooled, time-studied, costed, and were in production. Enough evidence of redundant and closely similar parts and assemblies, even with this costly and cumbersome procedure, prompted an executive edict to stop reinventing the wheel.

This, in turn, resulted in the classification and coding of all ingredient data, and excessive variety was prevented before rather than after the fact. Parts and assemblies are now represented by their supporting design data. They are grouped in families of similar shape or function in the computer. Anyone wishing to see the variety in a family can do so by inputting the classified characteristics of the data to find the family number. In this case, data visibility has been especially useful in curtailing unnecessary new parts designs, and for simplifying after-market variety.

This firm found that when an item was simplified and removed from the existing service parts file, it saved several thousand dollars for each item eliminated by combining it with another existing usable substitute. Data visibility retrieved in a logical sequence, then, is an essential first step in a simplification project.

Having the data visible, however, will not cause simplification to happen. You have to make it happen. It requires organization, goals, objectives, procedures, and a will to act. The benefits are well worth the effort. Returns on investment annually exceed the costs involved by a minimum of 100% per year and in some instances may reach 200–300% annually until the backlog of redundant items has been eliminated and variety controls are curtailing the creation of more redundancy. So regardless of the influencing factors of labor availability, project justification, funding procedures, and other deterrents to act upon the problem, it is very good business to simplify and standardize.

How, then, does one go about the simplification of an existing body of data? We can demonstrate this procedure by using an example taken from a preliminary feasibility study conducted for an European firm that manufactures construction machinery.

We examined procurements for 2 months out of 12. We found, for example, that 38 thicknesses of steel sheet and plate were purchased during the 2-month sample study period, in 131 different sizes by lengths and widths. (*Note:* When this firm classified, coded, and simplified their primary in-product materials, all of their active items were included.) This firm bought steel at three of their five plant locations. Of the 38 thicknesses found, 24 were metric sizes. The balance (14) were British Imperial sizes. Obviously, the 14 imperial sizes were made nonpreferred for future design use. However, some metric equivalent thickness had to be substituted—not size for size, of course—but of a sufficient number to enable the product to be made and perform to specification.

That 24 of the 38 thicknesses were expressed in metric measure is not sufficient reason to accept them as a preferred variety for future design. To determine what sizes would suffice for present and proposed needs, the British Standards Institute recommendations for metric sheet and plate thickness were examined.

There are 50 items listed in the steel plate and sheet recommendations (DDS: 1971). Of these, as many as five of the thicknesses have been noted as possible nonstandard items (i.e., one or more may be withdrawn). At the time of issue, no additions were contemplated. Thus, there will be 50 or fewer standard items eventually.

There are standards for basic preferred numbers based upon the geometric progressions developed by Charles Renard in 1879. These series are commonly designated by an "R" number to diffferentiate one progression series from another. The R5 series, for example, refers to a geometric progression in 60% incremental steps. The R10 series is in 25% increments, the R20 is in 12% increments, and the R40 is in 6% increments. All series are also suitably rounded. Figure 5.1 shows the ANSI Z17.1 standard.

5–SERIES (60% STEPS)	10–SERIES (25% STEPS)	20–SERIES (12% STEPS)	40–SERIES (6% STEPS)
10	10	10	10
			10.6
		11.2	11.2
			11.8
	12.5	12.5	12.5
			13.2
		14	14
			15
16	16	16	16
			17
		18	18
			19
	20	20	20
			21.2
		22.4	22.4
			23.6
25	25	25	25
			26.5
		28	28
			30
	31.5	31.5	31.5
			33.5
		35.5	35.5
			37.5
40	40	40	40
			42.5
		45	45
			47.5
	50	50	50
			53
		56	56
			60
63	63	63	63
			67
		71	71
			75
	80	80	80
			85
		90	90
			95

Figure 5.1 ANSI Standard Z17.1 for basic preferred numbers (decimal series). Preferred numbers below 10 are formed by dividing the numbers between 10 and 100 by 10, 100, and so on. Preferred numbers above 100 are formed by multiplying the numbers between 10 and 100 by 10, 100, and so on. (The percentages in the headings are approximate.)

Preferred numbers between 10 and 1 and below 1 are found by extrapolation. Each of the numbers between 10 and 100 is divided by 10, as in the following example. This is an extrapolation of the R5 series:

R5 for <10 to ≥1 (Preferred Numbers)	R5 for <1 to ≥0.1 (Preferred Numbers)
10 ÷ 10 = 1	10 ÷ 100 = 0.1
16 ÷ 10 = 1.6	16 ÷ 100 = 0.16
25 ÷ 10 = 2.5	25 ÷ 100 = 0.25
40 ÷ 10 = 4	40 ÷ 100 = 0.4
63 ÷ 10 = 6.3	63 ÷ 100 = 0.63

In the same series, for numbers greater than 100, it is necessary to multiply the numbers between 10 and 100 by 10. This would produce the following:

R5 for 100: (Preferred Numbers)
10 × 10 = 100
10 × 16 = 160
10 × 25 = 250
10 × 40 = 400
10 × 63 = 630

5.2 HOW TO USE PREFERRED NUMBERS

We compared the incremental differences on the British Standards Institute recommendations for steel sheet and plate and superimposed the metric sizes (thicknesses only) that were purchased by the British firm with an R10, an R20, and an R40 series.

In Figure 5.2 we show the R10 series only because this is what we would probably recommend. The comparisons are most revealing and instructive. The incremental percentages of the BSI recommendations do not conform to any rational geometric progression of numbers. Curiously, this is in contrast to BSI recommendations for wire sizes. An R10 series (Renard) was first choice in B4391:1972, with an R20 second choice, and R40 a third choice—a different committee and a different viewpoint!

Increment Increase	Thickness					
	mm	R10 (25%)	mm	R10 (25%)	mm	R10 (25%)
	0.50	0.50	12		65	
20%				12.5		
	0.60		15*		70	
17%		0.63				
	0.70		16*	16.0	75	
14%						
	0.80	0.80	18*		80	80
13%						
	0.90		20*	20.0	85	
12%						
	1.0	1.0	22*		90	
10%		1.3				
	1.6	1.6	25	25.0	95	
45%						
	2.0	2.0	28		100	100
25%				32.0		
	2.5	2.5	35		120	
25%						
	3.0	3.2	38		125	125
33%						
	4.0	4.0	40	40.0	130	
25%						
	5.0	5.0	45		135	
20%						
	6.0	6.3	50	50.0	140	
33%						
	8.0	8.0	55		145	
25%						
	10	10.0	60		150	
20%				63.0		
	12					160
					51	26

* In this area there is insufficient evidence to determine which of these thicknesses may become standard. Certain may be withdrawn with no additions.

Figure 5.2 Comparing standard thicknesses for steel plate and sheet with an R10 series: British Standards Institute recommendations (DDS: 1971).

Using the 2-month-sample-period figure, we pointed out the simplification potential for reducing variety in thicknesses alone. In Figure 5.3 the potential reduction in thicknesses reflected in the items actually being bought would be 28.4% of that number, and after simplification of widths and lengths, the figure would be greater still. In fact, raw materials this firm had been using prior to the classification and

From Exhibit E-1, you can see that you purchased metric sheet/plate sizes as follows:

1.	5 mm	9.	13 mm[a]	17.	45 mm[c]
2.	6 mm	10.	15 mm[b,c]	18.	46.6 mm[a,c]
3.	7 mm[a,c]	11.	16 mm[b]	19.	50 mm
4.	8 mm	12.	19 mm[a,c]	20.	60 mm (63)
5.	9 mm[a,c]	13.	20 mm[b]	21.	65 mm[a,c]
6.	9.5 mm[a,c]	14.	25 mm	22.	70 mm[c]
7.	10 mm	15.	30 mm (32)	23.	80 mm
8.	12 mm	16.	40 mm	24.	90 mm[c]

[a]Nonpreferred BSI.

[b]May be withdrawn by BSI.

[c]Not to R10 series.

Of the 24 metric thicknesses that you bought during the 2-month sample period, seven thicknesses are not on the list of recommended sizes published by the British Standards Institute (29.2%). Of the 88 metric items that you bought in these thicknesses, 25 (28.4%) are not preferred and undoubtedly will not be to a BSI standard. Thirty-three percent are not to an R10 series. When metallurgical specifications, lengths and widths, and chemistries are also simplified, the number of nonpreferred items and the percentage of the total will be higher.

Figure 5.3 Simplification potential in study report.

coding of the data have been reduced by 54.7% of the former variety. This is a net reduction after additions to the data base of some 14% new metric sizes. This was done to provide an adequate substitution for all British Imperial measured items, which had been declared nonpreferred for future design use.

5.3 STEPS IN SIMPLIFYING A POPULATION OF DATA

The steps to be taken in simplification of an existing population of data are as follows:

1. Identify and classify all data into coded familes.
2. Make parameter plots of all data by family where there are 20 or more items in a family. The parameters chosen must be relevant to the item population.
3. Apply an agreed-upon series of preferred numbers; or, for

characteristics not subject to preferred numbers application, use industry or institutional guidances; and/or good common sense, based on your experience with your firm's needs.

4. Indicate preference for future design use, by some signal code designator or other procedural means.
5. Establish disposition procedures and guidances for all items other than preferred items.
6. Establish a simplification project team to determine the most effective utilization of the surplus variety of items declared nonpreferred.
7. After sufficient experience with the simplified variety—say, 12 to 18 months—amending as necessary during this period, begin to establish standards.

To assure a practical conclusion of step 7 and make it a reality, it is imperative that variety controls be in place and users be in compliance with the preferences.

5.4 SIMPLIFYING STANDARDS

No, this isn't a misprint. Simplification has been stressed as a step prior to standardization, and it *should* precede standardization—that is, when establishing the standards. The problem referred to here concerns design and engineering standards that are already in place. All too often, standards procedures are such that:

1. They embrace an excessive variety or permit the proliferation of variety while trying to cater for every possible eventuality;
2. They cater for performance without regard for variety control; and/or
3. There is no quick response method to update (i.e., add, delete, or amend) the items contained.

A reasonable* test of standards and their capability to restrict unnecessary variety is to determine what percentage of the standard items has ever been selected for design application, according to the following appraisal:

1. 0—30% = poor

*This is not an infallible measure, as some firms list all items they have used, regardless of essential variety, which is also a poor practice.

2. 31–60% = fair
3. 61–80% = good
4. 81–100% = excellent

If your experience is like most, the majority of your standards will be in either category 1 or 2. Less than 3% of firms with standards programs will have an excellent rating of 81% or better.

One example involves a firm that is very proud of its standards. They have assigned standard numbers to almost 500,000 items, of which fewer than 95,000 have ever been used by any of their divisions and/or plants within divisions in the 20 years or more that they have been published. Another firm has among its many standards one that permutates to yield in excess of 125,000,000 different possible standard parts.

The two firms are both excellent competitors in their respective industries and their products are in great demand. The problem is that in both instances, the standards function is passive. There are no controls in place to monitor compliance—compliance is voluntary. There are no means to audit the effectiveness of the standards to control variety items to reflect actual needs.

With variety available in these two sets of standard parts, it hardly seems possible that similar but nonstandard items could exist. But such was the case. Hundreds of similar but nonstandard parts were found in the populations when their data were classified and coded. About the only redeeming feature of these standards sets are that they are already 90% obsolescent or obsolete. Virtually all are expressed in U.S. Customary units.

An example of simplifying existing standards can be demonstrated by examining the extensive variety of materials found in a firm in the machine tool industry. This manufacturer had two subsidiaries in different geographic locations. It was decided to consolidate all manufacturing in the parent firm's main plant location. Each of the subsidiary firms had its own engineering and part numbering systems. This meant that there could be three (or more) numbers for items in common use—even standard items.

During the course of a program to overcome this and related problems of identification and rationalization of common materials, it was found that a considerable variety of purchased components were closely similar, although different. Size-range plots were made after classifying and coding the items. A typical example involves commercial and precision tapered machine pins. There were 8 precision-type items and 97 items of the commercial type in use. The following simplification preference recommendations were made and accepted:

A. Commercial-type pins only were made preferred.
B. Commercial-type pins not to ANSI Standards were made nonstandard and were not to be used even on current designs—with no exceptions. Engineering changes were issued for design changes to accommodate preferred sized pins. Residual stock was transferred for service use only.
C. Sizes 2/0, 1, 3, 5, and 7 were made nonpreferred for future design.
D. Lengths ⅞, 3¼, 3¾, 4¼, 4¾, 5¼, and 5¾ were also nonpreferred for future design.

The preferred population remaining was 31 sizes, as compared to 105 sizes prior to simplification—a reduction of 74 sizes, or 70.5% of the former variety. Twenty-two months later, most of the preferred items became company standards.

5.5 THE OPPORTUNITY THAT METRICATION OFFERS

The United States is in an eviable position. Metrication offers an unparalleled opportunity, literally, to be reborn as a dominant competitor in world markets. Unfortunately, some might not think so if the negative remarks of some managers were heard. For example, the standards engineering representative of a firm in the automotive industry at a conference on metrication held in 1974 stated his firm's position on metrication and its schedule for changeover. To quote the firm's chairman and chief executive officer: "This firm's policy regarding metrication is that we will be the last manufacturer in the United States to metricate."

This statement was in sharp contrast to the response given by a Caterpillar Tractor Company representative. "Cat" converted to an all-metric mode several years previously. In fact, they had even soft-converted drawings from U.S. Customary measure to metric measure to simplify shop problems with a dual-measure language—gages, tools, and the like.

General Motors has had a top-level executive speaking to interested groups on the subject of metrication for several years. The message in his speech (and its principal thrust) is that metrication is the biggest non-event in General Motors history, or words to that effect.

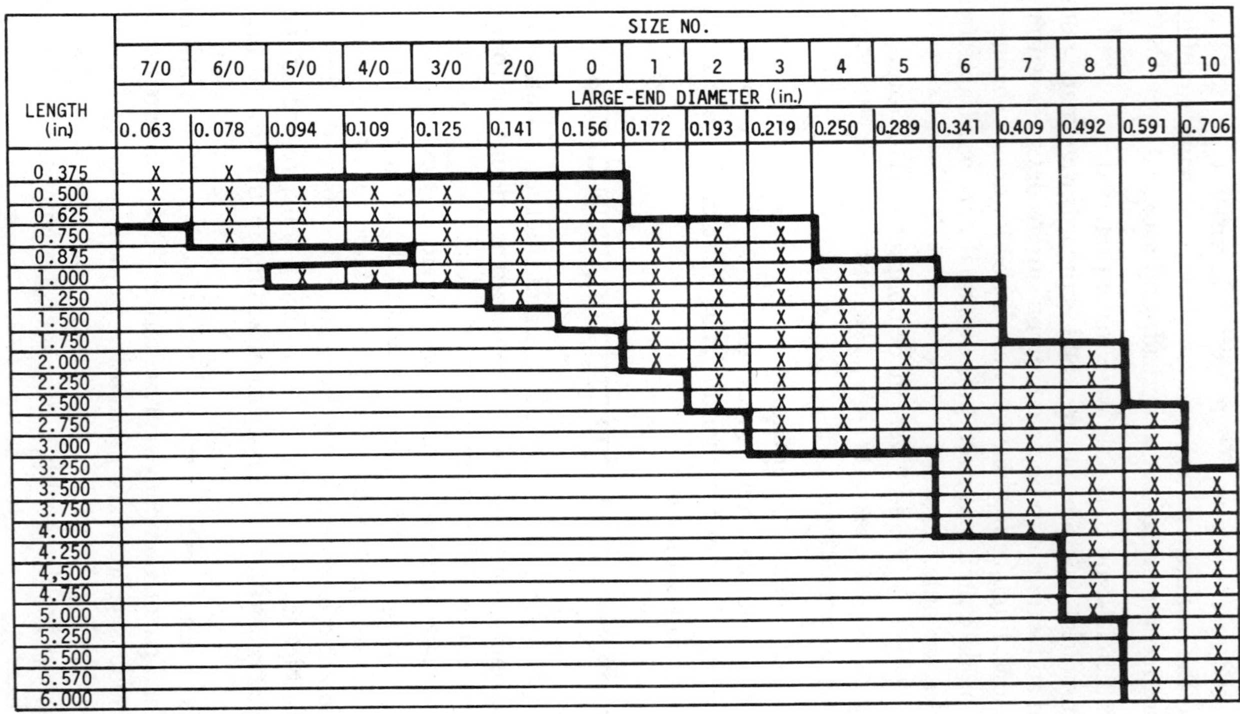

LENGTH (in.)	SIZE NO. 7/0	6/0	5/0	4/0	3/0	2/0	0	1	2	3	4	5	6	7	8	9	10
	LARGE-END DIAMETER (in.) 0.063	0.078	0.094	0.109	0.125	0.141	0.156	0.172	0.193	0.219	0.250	0.289	0.341	0.409	0.492	0.591	0.706
0.375	X	X															
0.500	X	X	X	X	X	X	X										
0.625	X	X	X	X	X	X	X										
0.750		X	X	X	X	X	X	X	X	X							
0.875					X	X	X	X	X	X							
1.000			X	X	X	X	X	X	X	X	X	X					
1.250						X	X	X	X	X	X	X	X				
1.500							X	X	X	X	X	X	X				
1.750								X	X	X	X	X	X				
2.000								X	X	X	X	X	X	X	X		
2.250									X	X	X	X	X	X	X		
2.500									X	X	X	X	X	X	X		
2.750										X	X	X	X	X	X	X	
3.000										X	X	X	X	X	X	X	
3.250													X	X	X	X	
3.500													X	X	X	X	X
3.750													X	X	X	X	X
4.000													X	X	X	X	X
4.250															X	X	X
4,500															X	X	X
4.750															X	X	X
5.000															X	X	X
5.250																X	X
5.500																X	X
5.570																X	X
6.000																X	X

Figure 5.4 Taper pins to ANSI B5.20 (137 sizes).

There are considerable cost-reduction opportunities for variety reduction with metrication. Examination of ISO/DIN Standards confirms this. Although these standards are comprehensive and, for the most part, very well conceived, the variety contained is excessive and not really necessary for most firms' needs.

We conducted a research project in this regard 4 years ago. The results are most interesting and instructive, and prove the point. The study was conducted by B.J.M. Lawrence, vice president and technical director for research and development for Brisch, Birn & Partners. Lawrence selected standard tapered machine pins as displayed in ANSI Standard B5.20 (see Figure 5.4). Currently, there are 137 inch sizes, ranging from No. 7/0 × ⅜" to No. 10 × 6". These pins are measured by large-end diameter and length, with a 1 in. in 48 in. pitch taper. When compared to DIN1 Standards (metric, of course) for taper pins (see Figure 5.5), it can be seen that there are 277 metric items as compared to 137 ANSI inch-sized items.

To make a realistic comparison, Lawrence converted the ANSI Standard pin-inch-measured dimensions to metric and used the small-end diameters as used in DIN1. Figure 5.6 reveals that 91 DIN sizes could replace the 137 sizes (ignoring the slight difference in the taper pitch). This is a comparison of two existing populations, however, and does not actually answer the question of control of proliferation of metric sizes before designs are created and are put into production at any one firm's facility.

One of the methods for simplification referred to earlier is to apply ANSI Standard Z17.1 for preferred numbers series. Which of the several series to use is governed by:

That which is closest to the existing population variety
The type of product your firm manufactures and/or assembles.

The second factor influences the limits of sizes likely to be used in your firm's products.

The diameters listed in DIN1 adhere closely to an R10 series and the lengths are a very close approximation of an R20 series (see Figure 5.7). The check marks on diameters and lengths indicate those items virtually corresponding to these two series (R10 and R20), respectively.

Because the present range of diameters approximates R10, to simplify these further dictates that an R5 series be used, which yields the following six diameters (X axis):

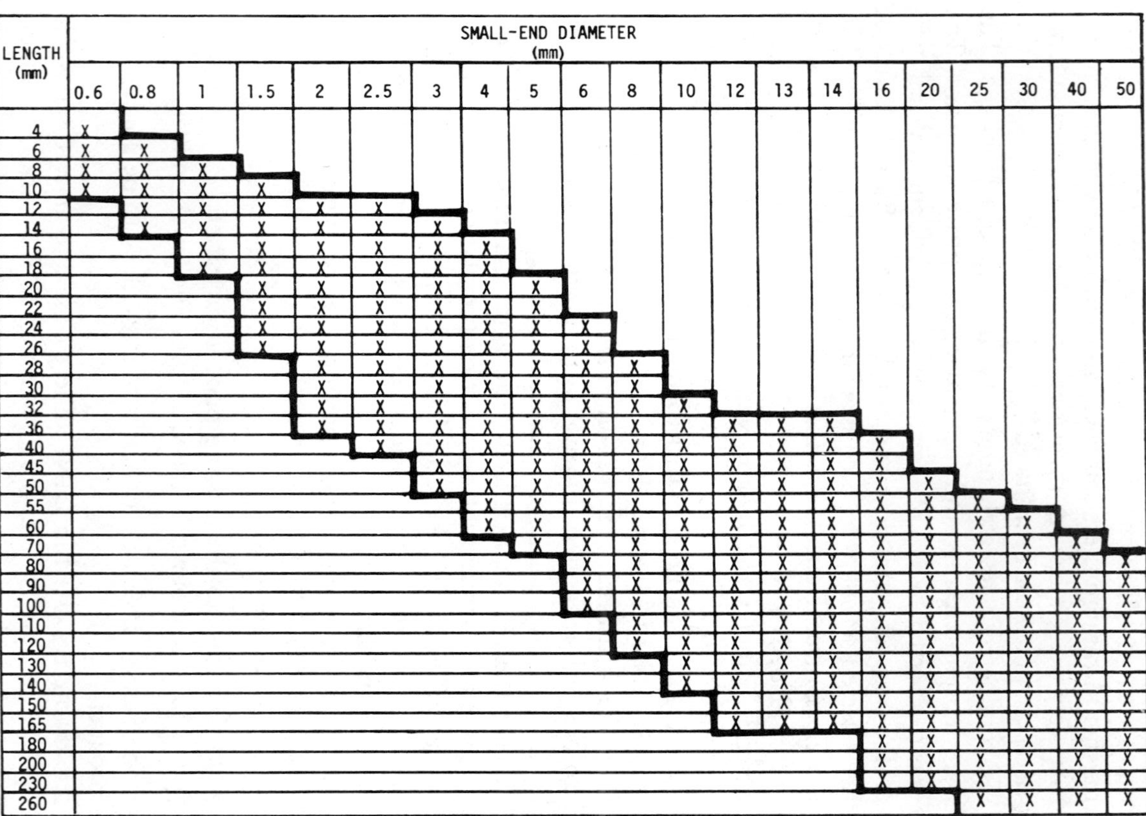

LENGTH (mm)	SMALL-END DIAMETER (mm)																				
	0.6	0.8	1	1.5	2	2.5	3	4	5	6	8	10	12	13	14	16	20	25	30	40	50
4	X																				
6	X	X																			
8	X	X	X																		
10	X	X	X	X																	
12		X	X	X	X	X															
14		X	X	X	X	X	X														
16			X	X	X	X	X	X													
18			X	X	X	X	X	X													
20				X	X	X	X	X	X												
22				X	X	X	X	X	X												
24				X	X	X	X	X	X	X											
26				X	X	X	X	X	X	X											
28					X	X	X	X	X	X	X										
30					X	X	X	X	X	X	X										
32					X	X	X	X	X	X	X	X									
36					X	X	X	X	X	X	X	X	X	X	X						
40						X	X	X	X	X	X	X	X	X	X	X					
45							X	X	X	X	X	X	X	X	X	X					
50							X	X	X	X	X	X	X	X	X	X	X				
55								X	X	X	X	X	X	X	X	X	X	X			
60								X	X	X	X	X	X	X	X	X	X	X	X		
70									X	X	X	X	X	X	X	X	X	X	X	X	
80										X	X	X	X	X	X	X	X	X	X	X	X
90										X	X	X	X	X	X	X	X	X	X	X	X
100										X	X	X	X	X	X	X	X	X	X	X	X
110											X	X	X	X	X	X	X	X	X	X	X
120											X	X	X	X	X	X	X	X	X	X	X
130												X	X	X	X	X	X	X	X	X	X
140												X	X	X	X	X	X	X	X	X	X
150													X	X	X	X	X	X	X	X	X
165													X	X	X	X	X	X	X	X	X
180																X	X	X	X	X	X
200																X	X	X	X	X	X
230																X	X	X	X	X	X
260																		X	X	X	X

Figure 5.5 Taper pins to DIN1 (277 sizes).

DIN SIZE		ANSI SIZE				
DIA. (mm)	LENGTH (mm)	SIZE NO.	mm DIA.	mm LENGTH	in. DIA.	in. LENGTH
1	8					
	10					
	12					
	14					
	16					
	18					
1.5	10	7/0	1.39	9.5	.0547	.375
	12	7/0	1.32	12.7	.0521	.500
	12	6/0	1.72	12.7	.0676	.500
	16	7/0	1.26	15.9	.0495	.625
	16	6/0	1.65	15.9	.0650	.625
	18					
	20	6/0	1.58	19.1	.0624	.750
	22					
	24					
	26					
2	8					
	10	6/0	1.78	9.5	.0702	.375
	12	5/0	2.12	12.7	.0836	.500
	14					
	16	5/0	2.06	15.9	.0810	.625
	18					
	20	5/0	1.99	19.1	.0784	.750
	22					
	26	5/0	1.86	25.4	.0732	1.000
	26	4/0	2.24	25.4	.0882	1.000
	28					
	·30					
2.5	10					
	12	4/0	2.50	12.7	.0986	.500
	14					
	16	4/0	2.44	15.9	.0960	.625
	18					
	20	4/0	2.37	19.1	.0934	.750
2.5	22	3/0	2.71	22.2	.1068	.875
	24					
	26	3/0	2.66	25.4	.1042	1.000
	28					
	30					
	32					
	36					
	40					
3	10					
	12	3/0	2.91	12.7	.1146	.500
	12	2/0	3.32	12.7	.1306	.500
	14					
	16	3/0	2.84	15.4	.1120	.625
	16	2/0	3.25	15.4	.1280	.625
	18					
	20	3/0	2.78	19.1	.1094	.750
	20	2/0	3.19	19.1	.1254	.750
	22	2/0	3.12	22.2	.1228	.875
	24					
	26	2/0	3.05	25.4	.1202	1.000
	26	0	3.43	25.4	.1352	1.000
	28					
	30					
	32	2/0	2.92	31.8	.1150	1.250
	36					
	40	0	3.17	38.1	.1248	1.500
	45	1	3.44	44.5	.1355	1.750
	50					
4	16	0	3.63	15.9	.1430	.625
	18					
	20	0	3.56	19.1	.1404	.750
	20	1	3.97	19.1	.1564	.750
	22	0	3.50	22.2	.1378	.875
	22	1	3.91	22.2	.1538	.875
	22	2	4.44	22.2	.1748	.875

Figure 5.6 Comparison of ANSI and DIN standards for taper pins.

X1 = 1.5mm
X2 = 2.5 mm
X3 = 4 mm
X4 = 6 mm
X5 = 10 mm
X6 = 16 mm

The range of lengths approximates the R20 series, which offers us a choice of R10 or R5 for simplification. Applying an R5, however, would limit the available lengths to just seven, as follows (Y axis):

Y1 = 10 mm
Y2 = 16 mm
Y3 = 26 mm

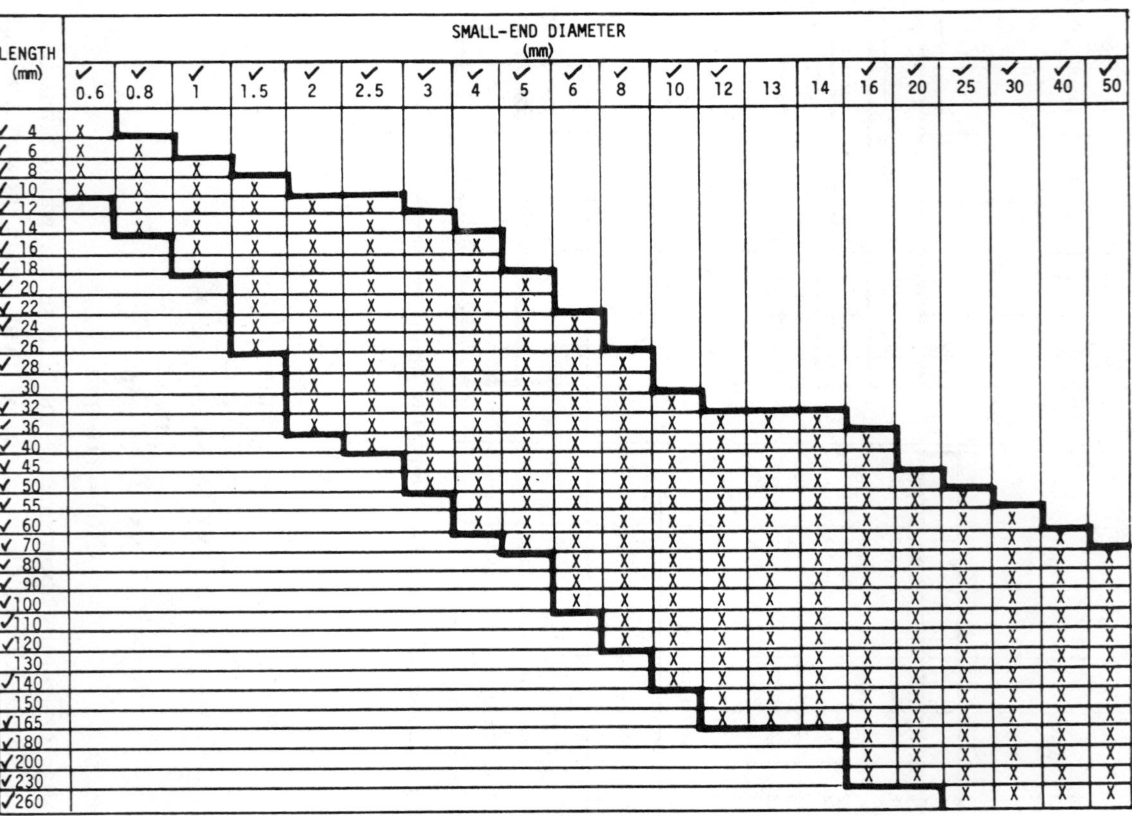

LENGTH (mm)	SMALL-END DIAMETER (mm)																				
	✓ 0.6	✓ 0.8	✓ 1	✓ 1.5	✓ 2	✓ 2.5	✓ 3	✓ 4	✓ 5	✓ 6	✓ 8	✓ 10	✓ 12	13	14	✓ 16	✓ 20	✓ 25	✓ 30	✓ 40	✓ 50
✓ 4	X																				
✓ 6	X	X																			
✓ 8	X	X	X																		
✓ 10	X	X	X	X																	
✓ 12		X	X	X	X	X															
✓ 14		X	X	X	X	X	X														
✓ 16			X	X	X	X	X	X													
✓ 18			X	X	X	X	X	X													
✓ 20				X	X	X	X	X	X												
✓ 22				X	X	X	X	X	X												
✓ 24				X	X	X	X	X	X	X											
26				X	X	X	X	X	X	X											
✓ 28					X	X	X	X	X	X	X										
30					X	X	X	X	X	X	X										
✓ 32					X	X	X	X	X	X	X	X									
✓ 36					X	X	X	X	X	X	X	X	X	X	X						
✓ 40						X	X	X	X	X	X	X	X	X	X	X					
✓ 45							X	X	X	X	X	X	X	X	X	X					
✓ 50							X	X	X	X	X	X	X	X	X	X	X				
✓ 55								X	X	X	X	X	X	X	X	X	X	X			
✓ 60								X	X	X	X	X	X	X	X	X	X	X	X		
✓ 70									X	X	X	X	X	X	X	X	X	X	X	X	
✓ 80										X	X	X	X	X	X	X	X	X	X	X	X
✓ 90										X	X	X	X	X	X	X	X	X	X	X	X
✓ 100										X	X	X	X	X	X	X	X	X	X	X	X
✓ 110											X	X	X	X	X	X	X	X	X	X	X
✓ 120											X	X	X	X	X	X	X	X	X	X	X
130												X	X	X	X	X	X	X	X	X	X
✓ 140												X	X	X	X	X	X	X	X	X	X
150													X	X	X	X	X	X	X	X	X
✓ 165													X	X	X	X	X	X	X	X	X
✓ 180																X	X	X	X	X	X
✓ 200																X	X	X	X	X	X
✓ 230																X	X	X	X	X	X
✓ 260																		X	X	X	X

Figure 5.7 Pins to DIN1 (277 sizes) compared to preferred numbers' sizes.

Y4 = 40 mm
Y5 = 60 mm
Y6 = 100 mm
Y7 = 160 mm

At this juncture, Lawrence concluded that this series might well be too restrictive and too drastic a step from his experience with medium and heavy industries/manufacturers. He concluded that an R10 series would be more likely to satisfy designer needs. This provides us with a total of 13 lengths, as follows:

Y1 = 10 mm*
Y2 = 12 mm
Y3 = 16 mm*
Y4 = 20 mm
Y5 = 26 mm*
Y6 = 32 mm
Y7 = 40 mm*
Y8 = 50 mm
Y9 = 60 mm*
Y10 = 80 mm
Y11 = 100 mm*
Y12 = 120 mm
Y13 = 165 mm**

Those lengths indicated by an asterisk are the coincident sizes of the R5 series; the double asterisk indicates a difference of 5 mm longer length than the 160 mm of the same series.

By combining the R5 series diameters with the R10 series lengths, the number of that preferred metric taper pin sizes should total 30, as seen in Figure 5.8. The impact would be enormous—a reduction of 247 DIN1 sizes. But more important, Lawrence concluded that there would be a net reduction of current ANSI Standards (B5.20) of 107—a reduction of 78.1% of our present standard variety—while still satisfying the majority of users and their needs. This kind of thinking is reflected in the proposed U.S. Optimum Metric Fastener Standards. This stand should be applauded and supported despite some of the objections of our European and Asian trading partners.

There are two conclusions to be drawn from the Lawrence study.

1. Our national standards for metric items should be as lean and trim as is possible.

LENGTH (mm)	SMALL-END DIAMETER (mm)					
	1.5	2.5	4	6	10	12
10	x					
12	x	x				
16	x	x	x			
20	x	x	x			
26		x	x	x		
32			x	x		
40			x	x		
50			x	x	x	
60			x	x	x	
80				x	x	
100				x	x	x
120					x	x
165						x

Figure 5.8 DIN1/ISO variety taper pins that should be preferred for future designs.

2. The potential for cost improvement through reduced variety far exceeds the cost of conversion, as proved by those who have already metricated.

Viewed in the light of the Lawrence paper and the experience of those that have already bitten the bullet and metricated, there is every reason to accelerate the establishment of sound metric product standards to the maximum rate possible in the United States.

The Australians learned this the hard way when their exports to third-world nations began to drop off. The principal reason, they found, was a resistance to Imperial measure. They metricated in 2 years' time (save for domestic consumer items). With the balance-of-payments problems in this country, we might well find there are competitive problems in addition to the high cost of oil.

Dramatic opportunities for profit improvement exist if we metricate properly. Some of the fallout benefits of a contraction in standard parts (tapered machine pins, for example) are as follows:

Fewer drill sizes required
Fewer tapered reamers required
Fewer after-market parts to stock
Larger lot size purchases, with attendant reduced costs/unit
Fewer purchases annually

Reduced inventories/investment
Fewer items to manage
Improved service levels
Lowered setup costs

We simply must improve our productivity and reduce costs. Metrication, properly planned and managed, may well provide a means to do this.

6

Installing Data Classification and Coding

6.0 SOME PREREQUISITIES

Managing industrial data variety is relatively uncomplicated with data visibility, provided that:

1. There is a willingness to restrict data proliferation.
2. There are checks and balances to ensure that only necessary new data enter the system.

Analysis of 30 years' experience with installations that have produced exceptional results, some not so exceptional, and some with average to poor results, has provided a few insights. But only in recent years has there been a logical explanation for the range of reactions and results. Why does the tool produce extraordinary returns on the investment in one firm, whereas in another firm it barely pays for itself—well below its potential worth?

The problem apparently has little to do with the quality of the classification and code. In fact, some very successful older programs for managing variety continue to produce excellent results with classifications and codes that appear somewhat archaic in the light of advancements made over the years.

If quality of classification and coding work performed is not at the root of the problem, what is? The answer has several parts, as follows:

1. The motivational aspects of how the program was organized and the work actually carried out (i.e., participation).

2. How well the program objectives were communicated by management.
3. Management attitudes on the importance of the program, with appropriate policy statements and accompanying procedures to support the program.
4. Follow up on the program by management.
5. Results reporting, with appropriate management commentary.

All too frequently, once a program has become operative, it becomes just one more program. If its manager fails to report progress, management promotions, retirements, and attrition quickly dissipate the rationale of why the program was installed in the first place.

The one characteristic in common that has distinguished supersuccessful programs from the less successful is an interested, sponsoring top manager. William Jack, in a paper entitled, "Top Management Evaluates the Classification and Coding Program," delivered at the Second International Conference on Classification and Coding, in 1967, at the Waldorf-Astoria Hotel, New York, said:

> Classification and coding has ramifications and possibilities as little understood by top managers today as computers were ten or fifteen years ago. Some managers are aware of the existence of such a tool, but few are aware of its potential.

He went on to say:

> Classification and coding affects every major function in a business with the possible exception of sales, and even sales benefits by improved deliveries, lower costs, and better spare parts service. In purchasing, things are considerably simplified; in stocks control, a number means one thing and one thing only. Production control and methods planning have operation charts filed in the same sequence code number with like parts together with the 5" × 8" drawing; progress expeditors can identify things easily; Data processing has a constant significant part number for family information processing; tooling and rate fixing can combine like tooling and times for like components; accounting is simplified, i.e., stocktaking, costing, and exception reporting.
>
> Three requisities for success in classification and coding to control and manage data are evident:
>
> 1. You must have an overall plan with Board knowledge and approval;
> 2. You must attack critical areas first; and

3. A member of top management must be responsible for the project on a continuing basis.

Where some classification-based codes installed a quarter-century ago may appear crude and unpolished in the light of advances made in the past 30 years, those systems are so deeply ingrained in management's philosophy that people who use the systems do not question their usefulness and function.

In Chapter 11, case histories provide details of installations—the good and not so good—and how problems were resolved to make them superproductive.

6.1 RESISTANCE TO CHANGE

For the moment, we will examine the problems attendant upon installing any innovation, because the acceptance and use of a data variety management program is no different from any other innovative system.

There is a natural conservatism in most people when it come to changes that affect how they earn their living. In some it is reactionary, even militantly so. This should come as no surprise to managers and industrial engineers who have been involved in methods improvement work. It certainly should come as no shock to industrial relations and personnel managers who have had to deal with work actions, strikes, or grievances, over new technology that either caused production output to be increased or labor content to be reduced. Even managers have reacted to innovations such as the computer, and exhibited many of the same symptoms as the hourly worker on the shop floor whose work was changed due to automation.

In the manufacturing engineering and product engineering functions, much interest has been shown by management in computer-aided design (CAD) and computer-aided manufacturing (CAM). Volumes have been written and billions are being spent to reduce the time and cost to conceive a design and get it into production. Logically, it should follow that what hastens CAD/CAM to become a reality should be accepted quickly. Certainly, those intelligent people who fight lead times from product concept to product reality should be most receptive.

The problem is that logic has nothing to do with it. Although much is said and much is being spent for CAD/CAM, unless certain motivation principles and amenities are observed in the process, there will be sufficient resistance to the innovations proposed to sink them before they get a fair trial. This applies to all changes, innovative or not.

The dilemma this country faces is how to raise its rate of productivity (i.e., more bang for the buck). Each working American has to produce more for the same cost, or the same amount for less cost. When that happens, competition sees to it that prices drop and markets expand and more people—not fewer—are then needed to produce more goods and services to satisfy the expanded demand. Everyone knows that we must improve productivity, but for the past 10 years, this has not been happening. Why, in the world's most productive nation, with annual productivity improvement steadily diminishing, when opportunities to raise productivity—however small—present themselves, are they resisted? The answer appears simple. Management apparently has not done its job.

If there is one lesson that can be learned from Japanese management, it is their painstaking care to seek the participation of all people who will be affected by any proposed change before adopting the innovation. The late Serge A. Birn, an internationally respected management consultant and productivity specialist, traveled around the world seeking answers to productivity problems. His interest was catholic and not restricted to U.S. problems. What he found in Japan was most revealing. In an interview shortly before his death, he said the following:

> In Japan, the annual productivity improvement rate is more than three times that of the United States. Productivity there is almost a religion. It is no accident that an employee in a Japanese firm subordinates his own interest to those of his nation and his company. It happens partly because of ancient traditions, but it is management, sensitive to these traditions, that is the difference. Even allowing for tangible government encouragement to improve productivity, such as no capital gains taxes, and government [encouraged] joint research, it is the traditions observed by management that respect the individuals' feelings, recognize their needs, and seeks their participation in decisions regarding their jobs that makes the difference. It is unparalleled in the world. The voice of every person who is likely to be influenced by a change in method, organization, procedure, system, or process, is heard. The time in preparing for change may seem excessive, but this participation before the fact, speeds the

implementation and, even more importantly, the acceptance rate. It is a curious irony that the only remaining country which practices the Christian work ethic is a Shinto/Buddhist nation.

6.2 A CASE IN POINT

If your personal experience with such things follows those of this writer, it is not too difficult to find analogies. One such example comes to mind where the results have been realized in excess of those anticipated.

A multinational firm that makes small applicances, floor care equipment, and white goods needed a new model code for its products. For 20 years it had been discussed again and again, only to be dropped. Because of the organization's need for a worldwide information system, it was decided that new part and product codes were essential.

Traditionally, product codes are assigned at two principal design centers: one in the United States for the western hemisphere, and in their U.K. partially owned subsidiary for the eastern hemisphere. The problems of product identity and variety in the United States are different from those of Europe, Africa, Asia, and Australia, owing to foreign nations' different product acceptance requirements. For 9 months, this writer shuttled from the U.S. plant to the U.K. plant and back. Every interfacing functionary on both sides of the Atlantic who would use the product code was either interviewed, participated in meetings, or both.

A typical problem was who does what to whom, and who needs to know what. The product engineering activity area traditionally assigns new product codes, in close collaboration with marketing. At meetings on the product code, with both marketing and engineering representatives present, the manager of engineering would painstakingly describe what marketing had to have in the code. As soon as he had finished, the manager of marketing would then describe in great detail what engineering had to have in the code. This scenario was repeated time after time in group meetings with various other functionaries.

Code requirements ranged from a model code no longer than three digits because "warehouse handlers will make too many shipping errors," to one with as many as 16 digits, describing every aspect of the product, including, I think, the color of the eyes of the engineer who designed it.

Gradually, facts emerged from opinions, and emotion gave way to reason, but not before every document on which a product code was likely to appear had been examined and classified as to whether code meaning was necessary to arrive at a decision or to take action. From a stack of forms 23 in. high, fewer than two dozen pieces of paper emerged, whereby a decision, or an action to be taken, required meaningful information in the format of the code. When agreement on code format was reached—first in the United States and then in the United Kingdom with some slight changes—and finally signed off in the United States 9 months and 10 days had elapsed. The actual code design and development work accounted for *less* than 12 days of that time.

The code has been installed for 6 years and there has not been one problem of any consequence on either side of the ocean. After 20 years of avoiding the problem, it was solved in less than a year. But the reason for its successful introduction and use has nothing to do with any genius in code development (even though it *is* very clever). The reason it was accepted so quickly after installation was that every single person who would use it had a say on what they *thought* they needed. What was developed was what they *actually* needed, and they had participated in reaching the conclusion.

6.3 CREATING POSITIVE ATTITUDES

Managers talk about X and Y theories of superior-subordinate relationship; about how communication problems arise; how to cure them; and on and on. Yet, although they can intellectualize about these problems, when push comes to shove, they tend to react emotionally to changes in basic organizational relationships. This is because their attitudes have not changed. Unfortunately, changes in attitude are not easily effected. Attitude alteration can not be legislated, willed, coerced, nor wished. How, then, can an attitude be changed to accept a better way to do something? How can an innovation be presented to those who must use it or make it work, so that their attitudes will be positive?

In a programmed instruction course entitled, "Job Design for Motivation",* this point is treated in one of the sessions. The title of the

*R. N. Ford, S. A. Birn, and D. A. Barnes, "Job Design for Motivation," PI-702, Serge A. Birn Co., Louisville, Kentucky, 1978.

session is "How to Bring About Changes in Employee Behavior." The reference text says the following:

> After World War II, the communications approach to solving employee problems was easily the most usual one when problems of productivity, quality, performance, or strife, would occur. Companies, especially large ones, would publish magazines, newspapers, motion pictures and, as things got worse, they would shift from black and white motion pictures to expensive color. They hoped that this would somehow appeal to the employee and help improve their productivity; to get "zero defects"; or to perform better. These programs regularly resulted in little gain; and any gains were mostly short lived.
>
> Communicating directly with employees about production problems or quality problems did not have a lasting impact, so the human relations effort of the late forties and fifties took a different approach from the communications approach, which was aimed straight at the employees. This human relations effort aimed to get the supervisor interested in the problem to ultimately change his behavior. We call this the "TAB" approach.
>
> T stands for *T*raining. We get the supervisors together for seminars, sessions, or some kind of learning situation and train them. We show them the problems, explain what has been happening, show them what needs to happen, and give them a new view of their employees. The Training is aimed at changing the *A*ttitude of the supervisor so that his *B*ehavior, in relation to his or her employees, would change too.
>
> TAB: *T*raining undertaken to change the *A*ttitudes, to change the *B*ehavior. The hope was that this revised attitude of supervisors would lead them to handle their employees more effectively. This, too, did not have a very profound effect.
>
> Let us consider a better approach. We call it "SBA," where S stands for *S*ituation. If we have bad productivity, bad quality, bad performance, or strife among employees, what situational changes can we make among the employees? Can we change the job content or the relationship of people to each other, or the flow of the product, or the service? *S* makes a Situational change, so that *B*, Behavior has to change also. Then, A, *A*ttitude change will be induced easily.
>
> Using this approach, you will get an attitude change on the part of the employees "free," if you can just change the situation leading behavioral relations to change. This will naturally result in

an attitude change. To give a simple illustration, suppose that you have a poorly performing typing pool. You can introduce more supervision to this pool. You can attempt to improve the quality of the supervision through training to change either supervisory attitude or behavior. Neither brings the desired result.

Let us consider a better way. Let us ask, "What can we do with this typing pool to improve the situation specifically? What situational changes can be make among these typists that will change their behavior, which will, in turn, result in a better attitude of the typists to their work?"

It might mean, for example, that you may have to rearrange the floor and reassign the typists to specific executives or floors, or parts thereof, as their normal or primary customers. We are not denying that you can change supervisory attitudes in a classroom situation and, thus, enable them to go back and change the behavior of their employees to a certain degree. We are saying that, in our view, the training approach is much less powerful than a situational change, as a basic premise.

Why not get the supervisors and managers in a joint meeting and have them review the situation which is causing difficulty? Let them freely discuss what is going wrong with productivity, quality, performance, or other problems they may have. Talk to the supervisors, hold sessions to get them to lay out, on their own, some situational changes that could be initiated. For instance, how could we arrange the flow of this work differently?

Under this approach, notice that we are not, in any way, attacking the supervisor's style of management. We are not labeling them as Theory X or Theory Y followers. We are not saying that they lack sensitivity, or that they employ paternalistic patterns of treating the employees as if they were children.

We are saying: — Do not attack the supervisor or manager and do not get them feeling guilty and defeated. Don't expect them to "tune in" to change their own behavior just through training. Don't expect a religious type of managerial conversion overnight toward changing the situation among their employees first.

Let us now talk about Behavior. Now, something new is happening with training in the United States which is well worth your attention. It is called Behavior Modeling. Before the supervisors actually get on to the floor with the employees, how about a few dry runs in which they will practice in a classroom or other protected situation exactly what they are going to say to these employees about the situational changes which are about to occur? And how? This is Behavior Modeling. If you have someone who can help in

> the classroom (a good instructor), he will let you talk, play back a video tape or tape recording, let you hear yourself, let you try it again, let you improve it, and so on. That is Behavior Modeling, and it can be part of any supervisory training situation.
>
> When you have a production problem, a quality problem, a customer problem, things are bad. Then, go ahead and have a supervisory meeting to discuss the situation. Decide what situational changes should be made first. Then, concentrate on what behavioral changes must take place on your part as a supervisor. These should then be modeled before you implement situational changes. The change in behavior for the employees will almost surely bring about the desired change in Attitude.

Suppose that the example of the typing pool were instead a drafting and design group and suppose the objective is to create a desire to accept a new control procedure to improve the group's productivity. The thrust of the session is that to change attitudes requires that situations be changed. Changes in environment, organization, and responsibilities alter traditional situations. In a well-planned program for installing data control and management, the participation of all the people expected to use the new procedures can not be too easy to achieve. And most certainly, it cannot be left to be done as a tail-end exercise after all development work has been completed. It must be done concurrently with development.

In firms where all the users are close at hand, this approach can be used more easily than in firms where the prosective users are in various locations, geographically dispersed by miles, or even separated by an ocean or two.

It is more realistic to consider representative participation in the development phase of a program and plan to include, on a rotational basis, other users for short periods—say, 4 hours a week or even 1 day every 2–4 weeks. When distance makes this prohibitive, a prolonged involvement—say, 3 days to a week—may be realistic. But somehow or other, let everyone have his or her say during the program development phase. This is an imperative.

6.4 CREATIVITY AND CONTROL: COMPLEMENTARY OR CONTRADICTORY?

In Chapter 5 we referred to standardization as the quintessential ingredient to maximizing the utilization of resources. There is absolutely no rationale to support any alternative, because redundancy of effort reduces productivty and wastes resources. But there has always been a conflict of emotions within most designers and engineering managers. They can accept the need to constrain variety intellectually, but emotionally their creativity seeks expression. It is well to recognize this because it is inherent. A designer is an artist of sorts. In fact, industrial designers and architects pride themselves on their artistry. To create two things exactly alike is foreign to their self-defined role in life. It is somewhat the same for the engineer and draftsman, from the artistic viewpoint. If Pope Clement had instructed Michelangelo to reproduce precisely the Vatican's Sistine Chapel frescoes in the cathedral of Notre Dame, one could image the reaction. Similarly, if he had instructed Michelangelo precisely as to what colors to use, what size figures to paint, and in what order, the Sistine Chapel ceiling would doubtlessly have been repainted more than once since. This presumes, of course, that the temperamental Michelangelo would have accepted the commission in the first place.

So, how to cater for creativity within reasonable constraints dictated by practicality? It is axiomatic that when an engineer or designer is assigned to create a solution to a design problem, the person will feel that he or she can conceive an innovative and unique end result. At the same time, to do this requires that the state of the art be known and/or accessible for study before it can be improved upon. This is where the problem lies. In most instances, the data are inaccessible and, therefore, not visible. When they can be made visible, there can be little argument that if the existing state of design is perfectly acceptable as a proven solution, creating an alternative may well be counterproductive. In point of fact, it invariably is a waste. The problem is to approach this dichotomy with an understanding of how to motivate the designer:

To be creative when it is necessary, but
To be just as motivated to use what is available that will solve the design problem.

One way to do this can be illustrated by an existing situation. It is not uncommon and follows a pattern that has been repeated a number of times. A multidivisioned corporation centralizes its product engineering in a headquarters design center. Problems with design complexity and design redundancy caused them to undertake a program of design simplification and variety control. The collateral objectives were for the purpose of reducing manufacturing costs and costs attendant with product design and preparation for manufacture.

A classification and code was developed covering a total population of approximately 50,000 individual items. At the same time, a computer-aided drafting and design system was being installed for hydraulic and electrical schematics, layouts, and wiring harness designs. Both innovations reached the engineers, designers, and draftsmen at about the same time.

A training program combining expository sessions, programmed instruction, and laboratory exercises was created and presented to 300 product engineers, designers, draftsmen, and manufacturing engineers. Manufacturing engineers had also been on the development team as well as a complement of design and standards engineers.

The training sessions reflected a good retention rate of the course content. The questions by the participants were searching, relevant, and constructive. And yet, when day 1 of the implementation arrived, very few of the designers used the service. In fact, the rate of inquiry continued low, in spite of an unusually high rate of design prevention. About one out of every three retrievals resulted in finding a usable existing piece part design. Even at the low rate of use, the program paid for itself in less than 4 months. Instead of encouraging more product designers to make use of the service, inquiries from product designers remained basically unchanged. One bright feature was system use by other functionaries. Manufacturing engineers, for example, made increasing use of the classified and coded data.

Review of experience on other projects has revealed that it has usually required from 12 to 36 months in a voluntary program before full benefits are realized. In mandatory programs, where part number assignment is part of the procedure, the use rate has been higher and the time frame for full acceptance and use somewhat foreshortened.

Upon advice of a behavioral scientist, a survey was made of the attitudes of those who should use the system. Attitudes were found to range from neutral to negative in most of the prospective users because:

1. No policy statement was made concerning management's position on:
    a. Staffing of the design function and/or changes due to a possible lower work load, from successful use of the service.

   b. Encouragement to use the service through inclusion of individual use statistics in merit salary review and management development growth status review.
2. No explanation was given to group leaders and section heads prior to the implementation sessions.
3. Little or no participation by other than the few who were on the development team.

To neutralize the deficiencies in the original implementation, the psychologist recommended the following:

1. Management to issue a white paper on the subject of productivity improvement.
2. A public relations series, supporting the white paper to be planned, to touch upon the impact on every person of improved productivity. The effect of productivity improvement upon
   a. Product cost, product price
   b. Competiton, foreign and domestic
   c. Inflation
   d. Value of the dollar, and
   e. Trade deficits

   was to be quantified and related.
3. A policy to be written on productivity improvement and a means to measure and share the benefits included in the policy.
4. All members of management affected to be given training in the psychology of motivation, behavior, and how to improve attitudes. This training to take one of the following forms:
   a. Structured program instruction
   b. Expository lecture training by professional behavioral scientists
   c. Structured program instruction supported by periodic professional involvement

   (c is the most expensive but the most positive approach.)
5. Small autonomous groups to be established consisting of a cross section of persons interfacing with the new systems being introduced. Included in a typical group will be a manufacturing or industrial engineer, a tool designer or tooling engineer, six designers and/or draftsmen, and perhaps a standards engineer, supervisor, or member of engineering management.

6. Each group to examine the problems relating to system design, job design, and the interactive nature of the two and to examine families of parts and materials to offer recommendations on how to simplify design features and materials effectiveness, and their effect upon productivity.
7. Reports of their findings and recommendations to be issued by each group to every other group. The reports to include specific recommendations on what to do about:
   a. Present opportunities to improve
   b. Guidelines for preventing the problem from recurring.

   These recommendations also to include, where appropriate, organization of jobs, job redesign, job responsiblities, and/or whatever else is consided relevant.
8. An action plan for each recommendation, including who will do what, when, and where, to be circulated to top managers in the product engineering and manufacturing engineering function. Each group is to be responsible to prioritize their recommendations.
9. A review of other groups' reports to be made. Each group's top-priority item to be studied by every other group and a master priority prepared by each group, reflecting all the groups.
10. One group to be selected to analyze each group's overall priority lists to develop and publish a master list of the top five actions to be taken, how, and by whom.
11. Departmental managements, in collaboration with the groups and the personnel function, to develop a new procedure for merit review to include the use of innovative systems and equipment as a measure of growth and responsibility assumption.

It is believed that these steps will, for the most part, overcome the deficiencies in the original implementation.

Earlier, mention was made of a faster acceptance rate when part number asignment was made part of the control procedure. This seeming contradiction is not meant as opting to use the data retrieval code number as the data part number identifier. However, there is a situation change involved and, because part number assignment is involved, there is usually greater emphasis in communicating before the fact. Especially is this so when the code is *to be the part number.*

It is easy to conclude that even though there is no plan to change part numbering systems, establishing part number assignment or part number controls as part of the implementation procedure has merit as long as participation before the fact is not ignored.

6.5 REPORTING AND FEEDBACK

Experience has taught that visual reports (i.e., graphs and charts) are more effective than tabular and expository reports. The reporting method and the procedures to make it easy to operate actually combine tables and graphs. Styled somewhat on the order of reporting methods developed by Rose and Farr,* the recommended program effectiveness reports include targets and/or goals. Loosely interpreted, the targets are objectives and the reporting process an exception report.

6.5.1 Designed Items

In Figure 6.1, an example of a control chart showing total piece parts designed, in both tabular and graphical presentation mode, is shown. The advantage of this method of presentation is that not only is the current month's activity shown, but a year at a glance is also shown. In this way, monthly fluctuations, which tend to distort a trend, are automatically smoothed. This charting method is also versatile in that either moving annual totals or moving annual averages may be shown.

Verbiage accompanying such reports is limited to one sentence, not to exceed 20 words, as to why the trend (either MAT or MAA) is more than 10% off plan.

Figure 6.2 is another control chart, which shows the number of retrievals that prevented design data from entering the system as a percentage of the total design activity. Here the moving annual average is used as well as the monthly activity. The chart makes it possible to see the net effect of the system.

The target is based upon the anticipated design activity for the upcoming year. For example, a product line is to be updated. This will result in greater-than-usual activity. At this point, the activity that will result must be analyzed in light of what is expected. If the update is

*T. G. Rose and D. E. Farr, *Higher Management Controls,* McGraw-Hill, New York, 1957.

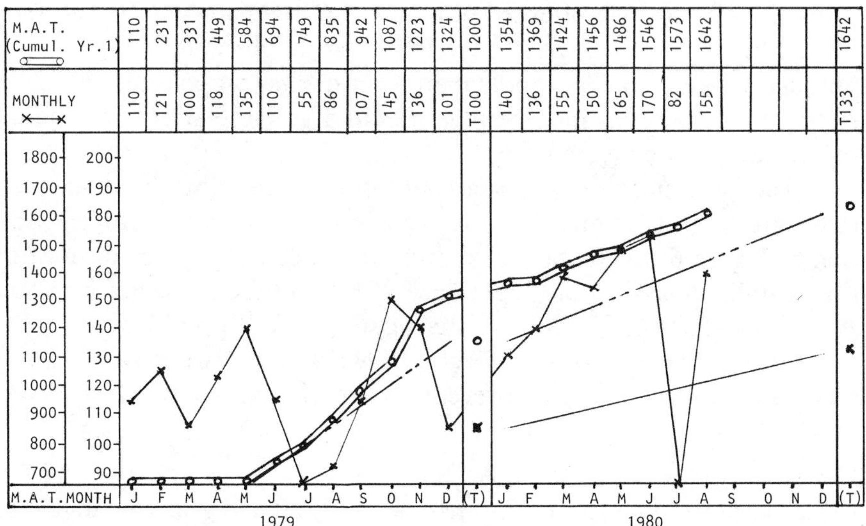

Figure 6.1 PIRC Control Chart 1, Class 3, showing the total parts designed.

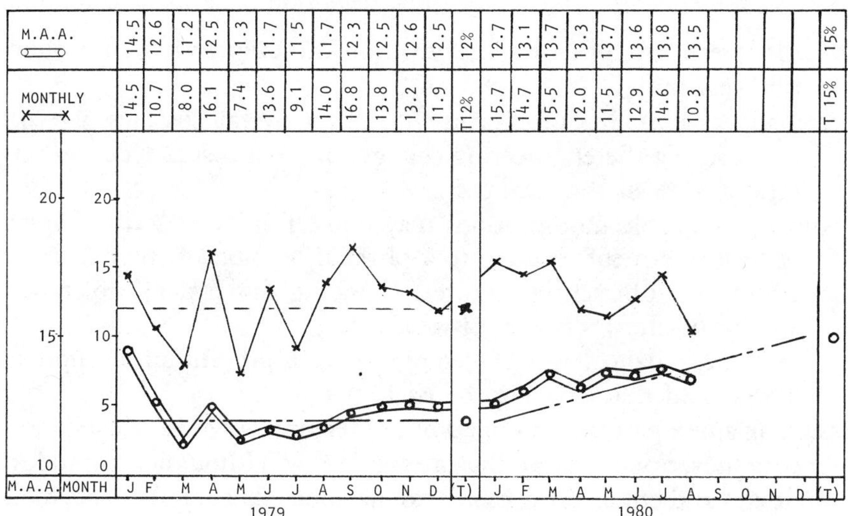

Figure 6.2 PIRC Control Chart 3, Class 3, showing the percent of new parts prevented by retrieval.

cosmetic only, or style-oriented only, the external parts will be affected. If, on the other hand, new materials make it possible to lighten the product, make it smaller, and so on, most of the components will need to be redesigned. So judgment is required and the situation studied before setting the target value.

The cycle from design data conception to entering the data identifier into the bill-of-materials file is shown in the graphical representation of Figure 6.3. The figures shown are about the average of most of the installations of our experience. The numbers in brackets are assemblies. The other numbers are for single piece part design.

Taking piece parts as an example, what this chart shows is that a reasonable average of design prevention over the first couple of years will be as follows:

Use existing design as is—10%

Use modification of an existing part without affecting interchangeability, but requiring an engineering change—3%

Use modification of an existing part affecting interchangeability requiring a new part number—6%

Make new design similar to, but different from, other members of the same family—81%

The relationship, in terms of cost-prevention benefits, is as follows:

Use as is saves 100% of design introduction and preparation for production costs.

Interchangeable modification saves 90% of the overall cost, less the cost of executing the engineering change, or a net cost prevention savings of 85% of the total cost.

Noninterchangeable modifications may range from a material change, or heat-treatment change, to a physical revision of the part (i.e., adding or subtracting features; changing dimensions, tolerances and/or finishes), or 50% of total cost.

On balance, the percentage of cost prevention is approximately half that of using the part as is.

Designing a new part that is similar but different saves 33% of the overall cost prevention. What this means is that although the product design and engineering cost is not avoided, there are benefits to all other affected users of the data visibility. This presumes that the installation is complete and being used in all departments.

As reported in Chapter 2, the cost to introduce a new part into produc-

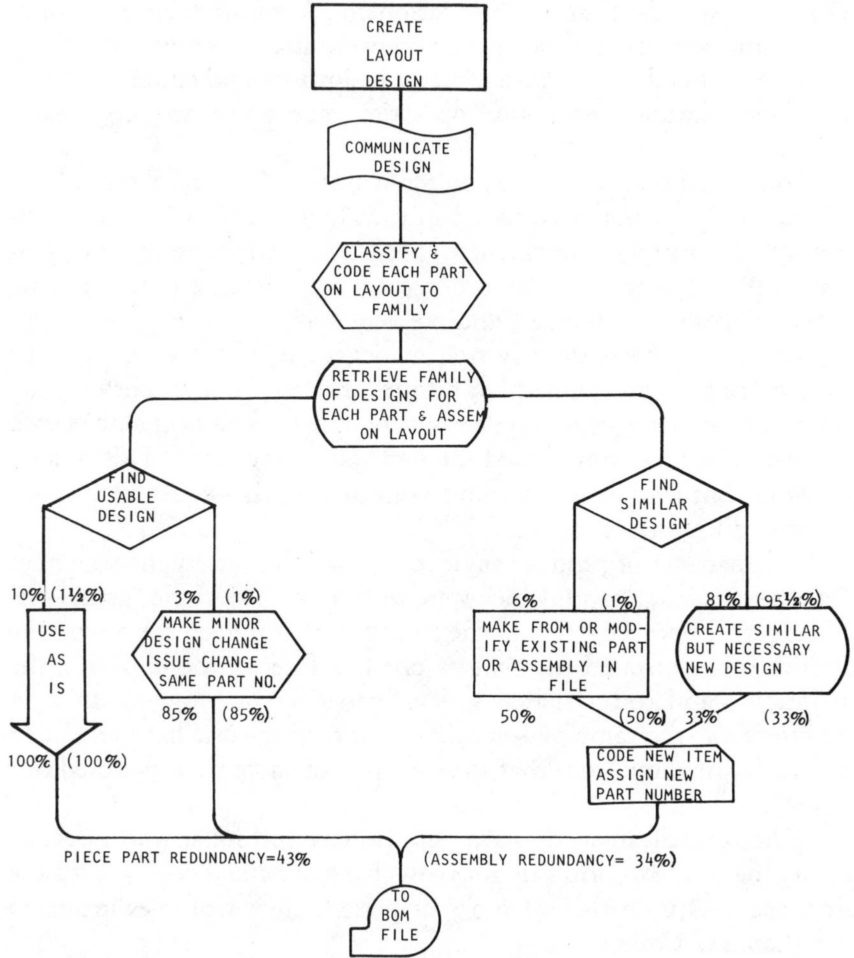

Figure 6.3 Design cycle.

tion will range greatly from firm to firm, but it is easy to see that if the weighted average is used, of the conditions reported, 43% of all costs to design and prepare a part for production can be avoided.

This has been verified by one firm that has documented savings on tooling alone. This firm makes products of sheet metal customized to meet design specifications of their customers. A considerable amount of their tooling consists of punch press and hydraulic press dies. The operations include blanking, forming, piecing, slotting, and cutoff. The dies are single operation, compound operation, and a few are progressive-stage dies.

The firm reports its savings on tooling only. That is, all other "cost to introduce" expense is excluded in calculating the return on investment of the variety management programs. All members of their management accept that the other costs are there, but the tooling figures are easily documented and are very real.

This firm is exceptionally well managed. It uses management by objectives as the basic game plan business strategy. Exceptions only are reported. Six years ago, tooling costs were only 60 + % of the objective. The board chairman questioned the low figures and said: "This is not a complaint, but is there an error in accounting, or did we really miss our budget and objectives by one-third?"

The manager of product engineering had the variety management program statistics at hand, which were within 2 or 3% of the "actual versus budget" difference. He told the chairman that indeed the reason the objective was overstated was that no one had forecast the impact of the classification and coding-based variety management program. In a recent interview, the same person said: "Our tooling costs have remained very nearly the same from that time in spite of more than doubled real volume".

What this means is that after adjustment for dollar worth decline due to inflation, this firm's tooling costs have remained constant for the past 5 years. During this same 5-year period, their real sales increased more than 2½ times.

Order quantities have increased as a result of increased application of parts data. The order repeat period has been shortened somewhat as well. The important fact to be remembered is that the nature of the business has not changed. This firm simply uses fewer different parts per customized product than before and there are more customers. It was explained that even where it was necessary to create a new part design, existing tooling could be modified where possible to produce the new items. When there was no way to change an existing die, it was possible to salvage punch and die sets, forming male and female inserts and the like to cope with the new configuration.

6.5.2 Nonproprietary Parts and Assemblies

The important figures for reporting inventory items are the record of items added and/or simplified out of the file. It is important to see the monthly activity as well as the cumulative effect. These figures are shown on Class 2 control charts, as follows:

1. Chart 1: Net status of file
   Items quantity.
2. Chart 2: Net movement of file
   Express as items change monthly and cumulatively.

A computer subroutine can provide the necessary data as follows:

1. Beginning number of items in file
2. Items added
3. Items deleted
4. Ending number of items in file
5. Difference

A typical example of Class 2 control would be the following:

1.	August 1	77,678
2.	Add	1,270
3.	Delete	6,033
4.	August 30	72,915
5.	Δ	(4,763)

This information is then posted on one trend sheet as shown in Figure 6.4(a and b). From these trend sheets, the data shown are then used to derive the charts shown in Figures 6.5 and 6.6.

Figure 6.5 is a histogram showing file size monthly. Figure 6.6 shows file change in quantitites of items, monthly and cumulatively. The data on Figure 6.6 may be extended to show as savings resulting from simplifying the inventory and by avoiding the cost of maintaining an item in inventory.

There are two kinds of savings that result from reducing present variety of these categories of material and preventing new items from entering the system. The first of these was treated in Chapter 2, which dealt with what the lack of variety control costs.

To refresh your memory, reducing nonproprietary raw materials, parts, and assemblies variety permits a freeing of inventory investment

DATE		K Items		
YR.	MTH	CURRENT	CUMULATIVE	M.A.T.
77	J	130.1	1.4	1.4
	F	128.7	2.3	3.7
	M	126.4	1.0	4.7
	A	125.4	.3	4.4
	M	125.7	3.5	7.9
	J	122.2	2.0	9.9
	J	120.2	1.5	11.4
	A	118.7	.2	11.2
	S	118.9	.4	11.6
	O	118.5	4.5	16.1
	N	114.0	2.0	18.1
	D	112.0	2.5	30.0
	(T)	100.1	2.0	20.1
78	J	110.0	12.3	32.4
	F	97.7	6.6	39.0
	M	91.1	3.4	42.4
	A	87.7	.3	42.1
	M	88.0	4.6	46.7
	J	83.4	5.7	52.4
	J	77.7	4.8	57.2
	A	72.9		
	S			
	O			
	M			
	D			
(A)	(T)	60.1	3.3	70.1

Figure 6.4 (A) Trend sheet: Class I PIRC Control Chart 1. (B) Work sheet: PIRC Control Chart 4.

capital for other uses. The relationship of variety reduction with investment-value reduction was illustrated by Figure 2.1. This is a one-time benefit. Figure 6.7 shows, for example, that if variety is reduced by 36%, it is possible to reduce the investment in the inventory resource by 12% of its overall worth.

In Chapter 2, reference was made to an annualized savings comprised of several factors:

Savings from avoiding the cost of carrying that portion of the inventory value now simplified out of the system.

Savings from fewer purchasing transactions because of buying more of fewer things.

Savings on item unit costs that result from quantity price breaks and/or, when applicable, pricing on total corporate usage versus individual divisioned quantities.

MONTH	1979 CURR. A Item	1979 CUM. B Item	1979 M. A. A. C %	1980 CURR. A Item	1980 CUM. B Item	1980 M. A. A. C %
J	16/110	16/110	A 14.54 B 14.54	22/140	172/1353	A 15.71 B 12.70
F	13/121	29/231	A 10.74 B 12.56	20/136	179/1369	A 14.71 B 13.08
M	8/100	37/331	A 8.00 B 11.18	24/155	195/1424	A 15.48 B 13.69
A	19/118	56/449	A 16.10 B 12.47	18/150	194/1456	A 12.00 B 13.32
M	10/135	66/584	A 7.41 B 11.30	19/165	203/1486	A 11.52 B 13.66
J	15/110	81/694	A 13.64 B 11.67	22/170	210/1546	A 12.94 B 13.58
J	5/55	86/749	A 9.09 B 11.48	12/82	217/1573	A 14.63 B 13.80
A	12/86	98/835	A 13.95 B 11.74	16/155	221/1642	A 10.32 B 13.46
S	18/107	116/942	A 16.82 B 12.31			
O	20/145	136/1087	A 13.79 B 12.51			
N	18/136	154/1223	A 13.24 B 12.59			
D	12/101 —	166/1223	A 11.88 B 12.54			

(B)

Figure 6.4 (continued)

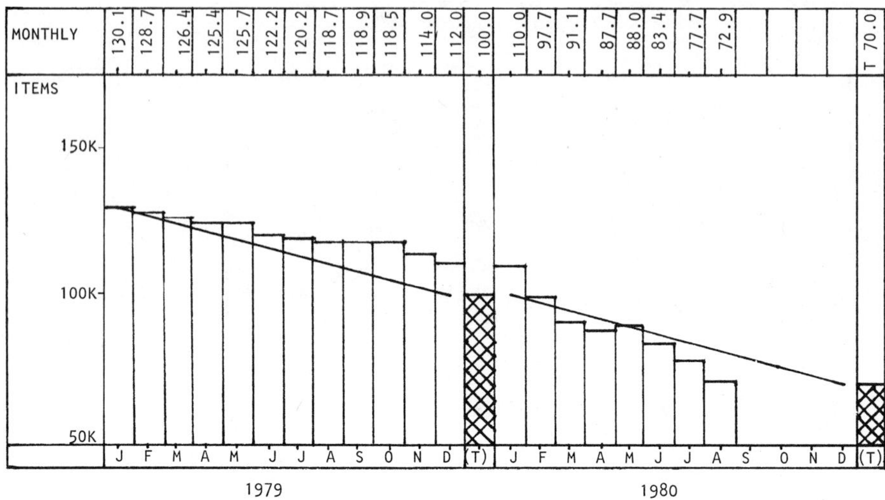

Figure 6.5 PIRC Control Chart 1, Class 2, showing a histogram of file size.

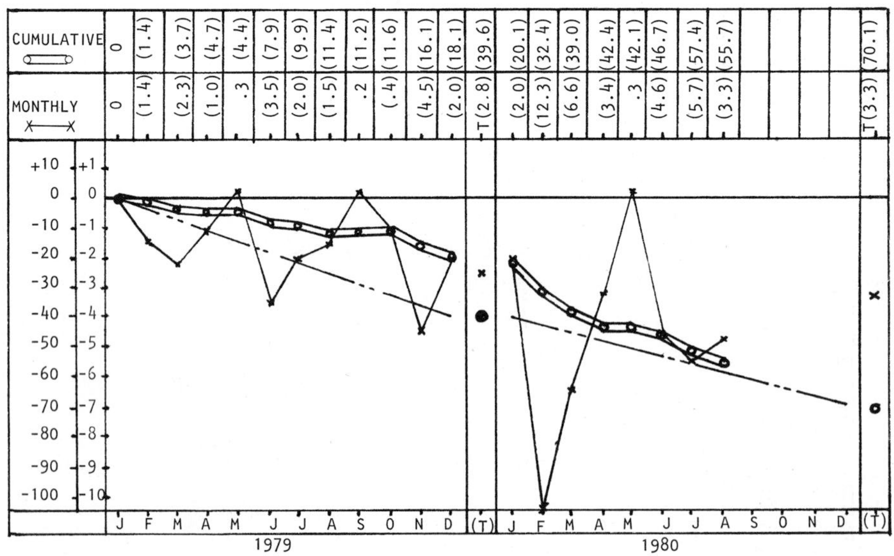

Figure 6.6 PIRC Control Chart 2, Class 2, showing file change monthly and cumulatively (thousands of items).

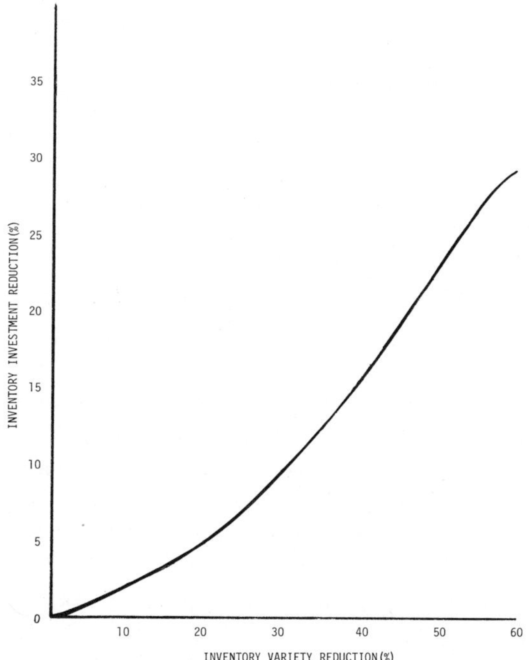

Figure 6.7 Inventory variety reduction versus inventory investment reduction.

For the purposes of survey on past practice, this is a satisfactory procedure.

For quantifying the impact of not adding a new item to the data file (i.e., using an existing preferred item rather than adding a new and probably nonpreferred item), a different procedure is required.

It is recognized that industry to industry, firms vary greatly in policy, organization, and practice. Then, what is applicable should be used. What is not applicable should be excluded.

The objective of the procedure that follows is to calculate what it costs to add a new item to the data file for the design life of the item. These costs are built up as follows:

1. Product by maker qualification and destruction analysis ______
2. Documentation necessary to procure part ______
3. Purchasing costs

   $______per purchasing transaction X______of purchases______

4.	Receiving inspection costs		______
5.	Other than 1 through 4, including:		
	a. Inventory management and administration	______	
	b. Housing and handling	______	
	c. Insurance and taxes	______	
	d. Deterioration, obsolescence, and mysterious disappearance	______	
	e. Manufacturing support, expedition	______	
	f. Interest charges per item	______	______
			======

One firm, manufacturing high-technology items, found an average life as an active item in the product of 5½ years with a cost of $2500, or $455 per item per year. This investigation was conducted by a task force comprised of five corporate vice presidents. [*Note:* Their interest charge on the value of the inventory (then around 9%) was not included. This is kept as a separate figure.]

To demonstrate how to use this procedure, the following figures will be used (with a 9% interest charge):

1.	Product by maker qualification and destruction analysis		$300
2.	Receiving inspection costs: 60 receipts × $10/receipt		600
3.	Documentation preparation		230
4.	Purchasing costs $24/transaction × 44 procurements		900
5.	Other than 1 through 4, including:		
	a. Inventory management and administration	$194	
	b. Housing and handling	32	
	c. Cataloging	15	
	d. Insurance and taxes	42	
	e. Manufacturing support	49	
	f. Shortages	44	
	g. Interest at 9% × 5½ years on average inventory value per item, $400*	180	
			556
			$2,586

*Reviewed quarterly (interest and average value of inventory per item).

Therefore, every time an inventory item can be prevented from entering the file, $2586 of cost can be avoided for the average 5½ years it will be active.

The problem is to record the use of existing preferred raw materials and nonproprietary parts and assemblies. To do this effectively requires that a program be written to extract the number from the file. According to conditions prevailing, it may be extracted from the bill-of-materials file, the where-used file, and/or by manipulation of the data base.

In the case of a manual operation, a manual procedure appended to the preparation of the bills of materials can be developed. Regardless of the source, the number of items prevented can be extended and plotted monthly and as a moving annual total.

Figure 6.8 differs from those showing file size, in that it deals with items prevented rather than file size movement. The file size is important to the realization of inventory investment reduction. The chart in Figure 6.8 is the savings to be realized from preventing the file to grow unnecessarily. It is not the effect of simplification, but rather the picture of how well ongoing variety is being managed.

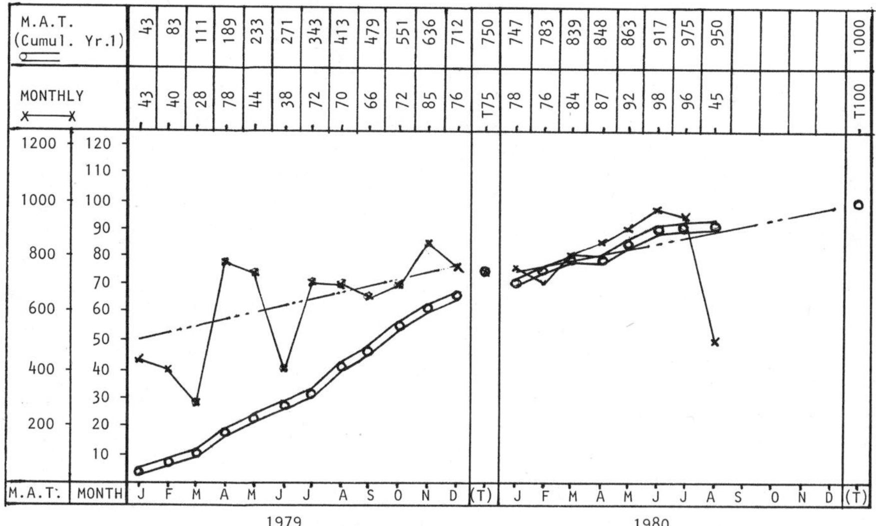

Figure 6.8 PIRC Control Chart 4, Class 2, showing new items prevented from entering the system.

In all candor, it is rare that this control is kept. The data contained should be disseminated to spur more use of the system and as a gage of the worth of the control procedure.

There are as many reporting systems as there are programs in use. Well-managed programs may differ in their method of reporting, but they all report to a senior member of executive management.

It is important to remember that there are large savings benefits to be realized *if* people are encouraged to participate, and top-management sponsorship has been found to be invaluable in this regard.

7

A Primer in Group Technology

7.0 GT: WHAT IT IS AND WHAT IT ISN'T

There has been much confusion as to what *group technology* (GT) is. At a 1969 conference on GT, sponsored by the International Labor Organisation at Turin, Italy, the most frequent subject of discussion was: What is group technology?* Granted that part of the problem was language (the speakers were English, Swedish, Russian, Czechoslovakian, Yugoslavian, Italian, and French) and differences in terminology, there were also basic conceptual differences. Essentially, there were four differing conceptual views, as follows:

1. The single machine concept.
2. The machine group or small cell concept.
3. The total organizational concept of GT — a variant of view 2 but requiring reorganization of policies, systems, procedures, and functions found in the line and staff of a firm.
4. The line flow concept.

The narrowest is the single machine concept. The broadest is a logical extension of the mass production "committed line" concept.

The confusion about group technology in the United States is even more pronounced. A good example of this confusion is the following. A

*J. L. Burbidge, *Proceedings of the Conference on Group Technology,* Turin Center on International Labor Organisation, Turin, Italy, 1969.

Delphi study was conducted by the University of Michigan in the mid-1970s; approximately 15 years after the first GT installation in the West (France). The title of the study was "Parts Production Technology from 1975 Through the Year 1990 in the United States." On the subject of group technology, the respondents projected that by 1979, group technology would be used on 25% of all manufacturing applications. Unfortunately, 80% of the respondents admitted to not knowing what GT is.

At present this lack of knowledge remains. A firm we know is embarked upon a program to improve product design standardization. They also plan to use group technology where applicable to improve costs by standardizing their manufacturing processes, increasing productivity, and reducing work-in-progress investment. Eight months into the program, a meeting was held to plan the implementation of classification, coding, and group technology. The working committee was polled on their respective definitions of group technology. The 14 members present responded as follows:

Three admitted they didn't know.

One said that GT is a means to define the function, geometry, and physical characteristics of an object (a definition of classification and coding, but not GT).

One defined GT as "using similarly classified parts to develop any technique by which these parts are made, purchased, sold, stocked, or dealt with."

Six defined GT as the standardization of the manufacturing processes and/or family scheduling — making similar parts the same way, scheduled in part similarity and process similarity order.

Three defined GT correctly, or at least distinguished cell manufacturing as part of their definition.

This group, which should have been better informed than most U.S. managers, showed very little improvement over the Delphi study respondents. Three mostly correct answers out of 14 is 21.4%—only slightly better than the situation obtaining 4 years earlier. Who or what is responsible for this confusion and continued ignorance about GT and what it is? Just about everyone who mouths the words and phrases but either has never read Mitrofanov's 1966 book* or bothered to get the

*S. P. Mitrofanov, *Scientific Principles of Group Technology*, 1966.

facts from those firms that have installed GT and/or have had success using the technique.

One person who has had a successful experience with GT is William Jack, Managing Director of JLG Limited. In the early 1960s, while financial director of Hopkinsons Ltd., Jack installed the first GT unit in the United Kingdom. At a seminar held by one of our most respected engineering societies, Jack opened his paper by saying: "Most managers and staff engineers pontificate on the subject of GT in much the same manner as a drunken person uses a lamp post—more for support than for illumination."

After his exposition of successful experience with classification, coding, and group technology, Jack added a note of warning and offered some sound advice for his American managerial counterparts. Commenting on the state of progress of GT in the United Kingdom, he made the following points rather strongly:

> Stop talking about GT and studying it endlessly. Get the help of someone who has had success with GT and install a pilot installation of your own. This way, you can learn by doing, correct your errors, and develop the feel for the technique in your own organization. To keep from destroying GT before it has a fair chance, *don't*—repeat, don't—follow the U.K. experience. Three things happened in our country that you [U.S. managers] should try to avoid, if at all possible.
>
> First, the pedagogues commenced to debate everything about GT, from what Mitrofanov really meant in his book to what classification and coding system(s) to use (if any), to the optimum size of a cell.
>
> Second, the Government got involved. Her Majesty's Government set up a special subministry [of Technology] section at Blacknest to promote GT in British industry. They muddied the water even more. Blacknest is now defunct after having squandered hundreds of thousands of pounds to no end.
>
> Third, generalist management consulting firms got involved and added the coup de grace. They were largely without firsthand experience in classification and coding. Their knowledge of GT was purely theoretical; gleaned from the combined output of the government and the college professors, most of which was contradictory, if not pure rubbish. After extensive promotion, they received a number of GT assignments, most

of which were done so poorly that U.K. managers became disenchanted.

These three things have indeed dampened the enthusiasm for the technique in the United Kingdom. This warning is very timely, as we appear to be repeating the same scenario in the United States.

The problem with definition can be resolved quite easily. The *single machine* concept is not GT but is useful and should be employed, regardless of batch-size considerations and/or layout. Specifically, this is the batch processing of similar parts in a family, one batch after the other, on the same machine(s). There can be no disagreement with this procedure. It is preferable, however, to call it "family scheduling" and/or "process standardization." Although related to GT, it did not originate with GT. It is conceded that to do this systematically rather than by accident coincided with GT development. But if GT had never come to pass, classification-based coded parts visibility enabled this realistic method of scheduling to become reality, and predated the introduction of GT in the West.

The *line flow,* or logical extension of mass production, is simply a variant of an existing product layout concept of committed tools. The principal difference between the two is that families of parts of similar process, regardless of the line of products they are used on, determines the candidates for loading the line, and batch size is small.

The most productive and most radical departure from the normal method of dealing with small lot batches is the second definition (the one on which Mitrofanov and the majority of experts concur). We have Anglicized the Russian definition of GT to read as follows:

> *GT* is a technique for manufacturing small to medium lot size batches of parts of similar process, of somewhat dissimilar materials, geometry and size, which are produced in a committed small cell of machines which have been grouped together physically, specifically tooled, and scheduled as a unit.

In the preceding definition are these imperatives:

Small to medium lot-size batches (i.e., one off to some experience-dictated restriction on order lot size).

Similar process [i.e., parts can be completed in the cell (except, perhaps, for heat treatment and/or surface treatment)].

Somewhat dissimilar materials, geometry, and size—within the size and shape constraints of the holding fixtures, chucks, clamps, and vises.

Processed in a committed small cell of machines, grouped geographically together—in process order or not—committed to processing only those candidate family parts, irrespective of machine utilization.

Specifically tooled (i.e., tool holders and/or tool stations remain fixed and are adjusted only—never removed!).

Scheduled as a group—the cell or GT unit is scheduled as if it were a single machine using a single scheduling cycle.

This definition does not exclude the total management (third) concept. It has worked very well in certain instances. Neither does it imply that the cell of machines cannot employ a line flow layout, provided however, that the handling labor savings can offset the expense of the handling equipment used. There is one important proviso on the latter. This does not imply progressive machining—part by part. The batch order is still processed in batch mode—operation to operation.

7.1 EVOLUTION OF GROUP TECHNOLOGY

Most who know group technology rightly acknowledge S. P. Mitrofanov of the USSR as the originator of the technique. But those who have studied the subject in depth agree that Mitrofanov's creative developments were founded upon a premise first postulated by A. P. Sokolovski (also of the USSR).

In 1937, Sokolovski suggested that parts of similar configuration and features—all other factors being equal—should be manufactured in the same way by a standardized technological process. Mitrofanov expanded upon this premise.

More specifically, while the process (machinery, tooling, and gagging) caters to the variety of sizes, shapes, features, and materials (within reasonable bounds, of course), the common denominator as to process unification (Mitrofanov's term) is the *method* of manufacturing the parts.

It is important, however, to note that Mitrofanov specifies design simplification and standardization as imperative prerequisities to an effective GT program. He specifically recommends a dual (multiplex) classification and code (i.e., one for design standardization and one for process standardization). In his paper presented at the GT Conference at Turin* was the following statement:

> Simplification and standardization of design make for technological variety reduction, with which the manufacture of many components can be carried out, using more modern and effective methods of an overall character common to all of them.
>
> In the case of individual components, or families of components, which are similar in design, highly productive methods are required, using common, easily adjustable tools. These methods make for a high level of productivity and earnings capacity for setup time and tool adjustment time.

He went on to say that the classification of technical operations based upon component shapes, surfaces, and features affords the best solution to this problem. He elaborated upon the role of classification as "*the basic problem [solution] on which Group Technology rests.*"

It is important to recognize that Mitrofanov, Zvonitsky, and others in the USSR (where there are more than 800 plants using GT) differentiate batching methods.

Mitrofanov refers to the unification of production process as consisting of the two types, as follows:

1. Standardization of the production process (irrespective of layout, process, and/or product).
2. Group production method (a committed cell of machines)

Another way of saying this is that for least-cost batching, there are two methods, either or both of which can be utilized:

1. Scheduling by family for grouping for manufacture on optimum standardized processes in process-oriented layouts.
2. GT cells of optimum standardized production processes.

Graphically, this can be shown as follows:

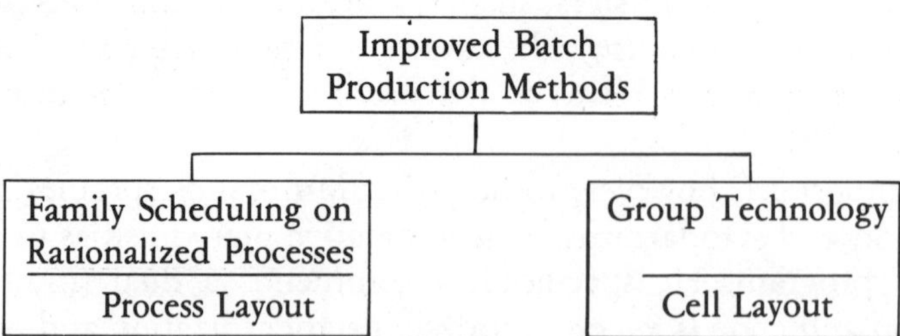

*S. P. Mitrofanov, *Proceedings of the Conference on Group Technology,*

It is interesting to note that a rudimentary form of GT was conceived, planned, installed, and operated more than 40 years ago in the United States. In 1937, the house organ of a U.S. manufacturer announced a new technique for making parts. It was in this same year that Sokolovski's paper was published.

The Cincinnati Milling Machine Company (now Cincinnati Milacron, Inc.) suffered the problems of low-volume batch production then, as now. To cut down the excessive flow distance and reduce production cycle time, their Oakley plant was arranged in cell layout mode. A rudimentary form of classification by cubage of the parts and/or noun descriptors was used to schedule parts to a cell set up to handle parts of that size or type. Granted that families of mill overarms, saddles, hand wheels, and the like helped in specifying tooling and fixturing, nonetheless, they grouped machines for low-volume batches and practiced a kind of "group technology" not too different from today's. Not only did the "Mill," as it is affectionately known to industrialists in southwestern Ohio, invent a form of GT, but they also conceived job enrichment 30 years before Herzberg knew what to call it. The story in their factory paper proclaimed that the workers of the future, at Oakley, would have to have more skills and know more!

The first planned group technology installation was made in 1958 at a French firm, Forges et Ateliers de Construction Électrique de Jeument. After a successful installation of a hierarchical monocode in 1955 to control in-product materials and components engineering data variety, it was decided to use the same classification in conjunction with group technology to resolve a serious batching problem in manufacturing. Lot sizes per order rarely exceeded 15 items with a 17-month frequency of repeat cycle.

The first production unit (cell) of two lathes, one vertical drill, and one milling machine became operational in 1958, drawing from 6000 candidate parts, with excellent results. Setup time was reduced by 85% and each-piece times were reduced by 20%. A second production unit was formed immediately, with equally successful results.

In 1961, a conference on GT was held in London. The results of the program at Forges et Ateliers were given. In 1963, the first GT installation was begun at Hopkinsons Ltd., in Huddersfield, England. This installation closely resembled that of the Forges et Ateliers installation set up several years earlier. In 1963, a suspect version of GT was installed at a machine tool manufacturing plant in Springfield, Vermont.

It was not until after 1966 that widespread attention was drawn to GT as a recognized manufacturing technique. It was this year that Mitrofanov's book was published first in Russian, then in German by the East Germans. Also in 1966, the First International Conference on Classification and Coding was held in the United States, in Fort Lauderdale, Florida. William Jack spoke on group technology at Hopkinsons Ltd. Joseph Gombinski, mananging director of E. G. Brisch and Partners Ltd. spoke on batching. In 1967, Jack and Gombinski spoke on GT and batching at the Second Conference on Classification, Coding, and Group Technology, held at New York's Waldorf-Astoria Hotel. In 1968, Gordon Ranson, then managing director of Serck Audco Ltd., Newport, England, spoke at the Third International Conference on Classification, Coding, and Group Technology, in Fort Lauderdale, Florida.

The attendance at these three conferences averaged 73 people each. Managers and industrial and manufacturing engineers were just not interested in GT. An editor of the then-leading American industrial management periodical, said the following in 1967:

> We know all about GT, but it is an obsolescent procedure. Numerically controlled machines, adaptive tooling and computer controlled machine technology have leapfrogged GT.

It is interesting that the editor who made that statement was out of a job in a few years as the venerable magazine he edited folded.

Today, thanks to organizations such as the Society of Manufacturing Engineers and a more enlightened industrial press, group technology is on practically every manufacturing engineer's agenda for investigation. For those who carried the word a dozen years ago to a disinterested America, it is most gratifying in this period of productivity crisis to see the awakening. Now, all that remains is to ensure that those who have become interested in GT know what it really is.

7.2 FAMILY GROUPING AND GT

Figure 7.1 has two parts. They have closely similar geometric configurations and similar features. It is evident that if these two parts are run in sequence, the tooling setup is a matter of tool adjustment, not resetting the machine. On a turret lathe, the resultant savings will be on the order of perhaps 2–3 hr. On an automatic screw machine setup, the reduction

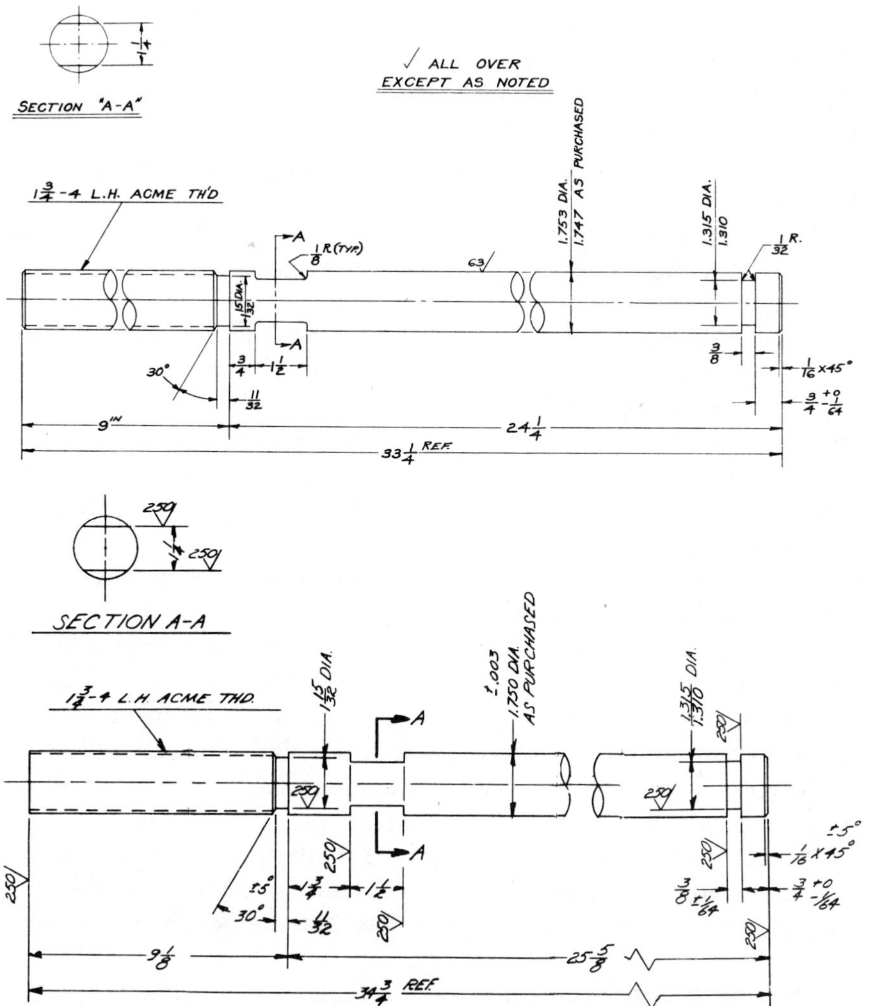

Figure 7.1 Similar parts for sequential family scheduling.

may well be 4–8 hr saved—or more. Even if there are 1000 or more parts per each of the two orders, three-fourths of a day of setup time and expense is of no small consequence. Therefore, lot size is not a qualification for using family scheduling or grouping. In GT, however, lot size *is* important and is one of the parameters for candidate part selection for processing in a GT cell.

The determination of which parts should be grouped together for GT cell production and which should be made in the conventional process-oriented shop is a decision for the production controller. It is important to keep the cell load on small lot orders. This is not always easy to do. There is constant pressure from sales and/or cost accounting and other functional departments to keep raising the lot-size acceptance parameters. This happens because of the vast difference in the overall production cycle. In the conventional shop, the cycle from receipt of raw stock, from stock preparation to the completion of the finished lot, may be as high as 1 week per operation. Seven operations means 7 weeks' flow time. In contrast, the GT cell production cycle rarely exceeds one shift—8 hr for the same number of operations.

It is evident that scheduling GT cells as individual units produces a marked reduction in work in progress inventory investment. This is primarily why the single machine concept of GT is fallacious. If the machines are not grouped together physically and not scheduled as a unit, there is little chance that work-in-progress (WIP) inventory will be materially reduced. Any reduction (2–4%) results from shorter cueing times because of reductions in setup time. Unless planning cycle times are actually reduced, there will be no reduction in the WIP investment. Without the reduction in work in progress, a major benefit to management is denied.

Another benefit of cell-arranged machines with small autonomous groups is that each piece time per part is reduced. The productivity of the GT cell laborpower is markedly higher than that of the traditionally organized shop. More will be said on this later in the chapter.

We can conclude from these arguments the following:

1. GT (cellular groups) should be used for small lot batches.
2. Family scheduling should be used for larger batches and all batches where it is impractical, if not impossible, to move the equipment into close-proximity cells.

7.3 ANATOMY OF A GT PILOT INSTALLATION

The best way to illustrate how to install a GT pilot is to track an actual experience. What was done, how it was done, the mistakes made, and how they were corrected may well be the most instructional method. Certainly living through the experiences at Forges et Ateliers and Hopkinsons provided our firm with numerous useful do's and don'ts for the future, one of which is to respect the individuality and unique circumstances each firm, or even each plant within the firm, exhibits.

The Hopkinsons Ltd. installation is the most memorable, perhaps because it was a first GT hands-on experience for this writer. What is imparted here is from interviews with the principals who were involved in the actual decisions prior to a firsthand involvement 2 years after the initial work had begun.

There is another good reason for using the Hopkinsons experience. It was one of only several of all the GT installations made where accurate before and after cost data were made available and formally reported. This was directly due to the fact that the entire project was personnally monitored by the authorizing director (vice president in the United States), who had first initiated the extensive classification and coding programs (tools, raw materials, commercial parts bought out, proprietary parts, assemblies, and products). The data were assured to be accurate inasmuch as William Jack was financial director—that's correct, financial director, of the firm. Their plants were operating at near capacity and space was at a premium. The manufacturing staff suggested making use of a very old plant some miles distant from the main plant. It had been used partly as a service parts warehouse and for light work associated with service packaging, With reluctance, all agreed.

The next problem was to find equipment. Lead waiting time for new equipment scheduled for the main shops was over one year (this was 1962–1963, remember). So a foraging expedition was organized to find an acceptable source for at least one turret lathe (called capstan lathe in the United Kingdom). In the meanwhile, the old building was being rearranged to make room for the then-unknown machinery.

Coinciding with these activities was the analysis to determine what parts were to be made once the building was ready and when machinery could be found.

Some arbitrary decisions were made as follows:

1. Lot-size restriction was 12, or fewer, parts per order (the Forges et Ateliers lot-size limit).
2. Ferrous metals only would be considered, to prevent switching from ferrous to nonferrous, and vice versa (this was later changed by an ingenious suggestion from one of the cell members for a removable tray to catch chips).

During the preparation period, the production planning group was asked to provide a list of all parts orders on which 20 or fewer parts were made, plus the annual repeat frequency (i.e., how many times during the year the part was ordered). Also requested was the actual hours charged against each lot for:

1. Setup
2. Each piece, if possible

The computer programs were written to process several files on which the data could be found. Most of what was requested was finally made available, but manually kept inventory records provided the bulk of the data—computer storage at that time was too restrictive and very expensive for keeping such information. Decks of cards were punched from the manual data records and the consolidated information data sheet prepared by the computer. (The programming costs were made part of the installation cost.) The family code number for each item was added from a cross-reference listing. In this way it would be determined which families (as ultimately were found) or which groups offered the best opportunities for GT for the pilot and follow-on work on other cells. Round parts with configurations such as those shown in Figure 7.2 were typical of what was eventually to be produced.

Working from the information on shape and functional data contained in the classification, the team assigned developed two composite components. These composites did not exist as actual parts, but represented the variety shapes and features of the parts. One composite shown in Figure 7.3 represents parts to be produced from bar stock through the head stock. The second composite represents parts that are either cast, forged, or require second operations.

Next: Process picture sheets were created that

1. Identified each surface and/or feature to be machined and the tooling required to create the configuration.

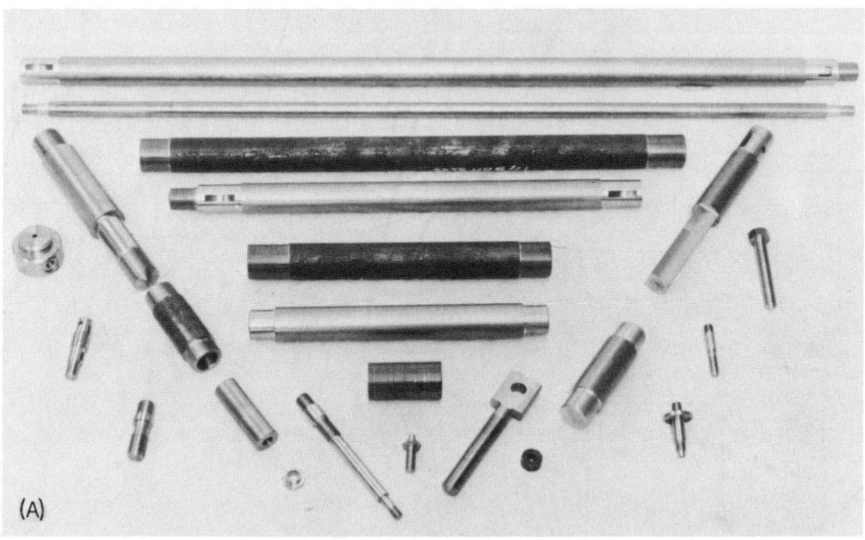

Figure 7.2 Round parts produced in a GT cell.

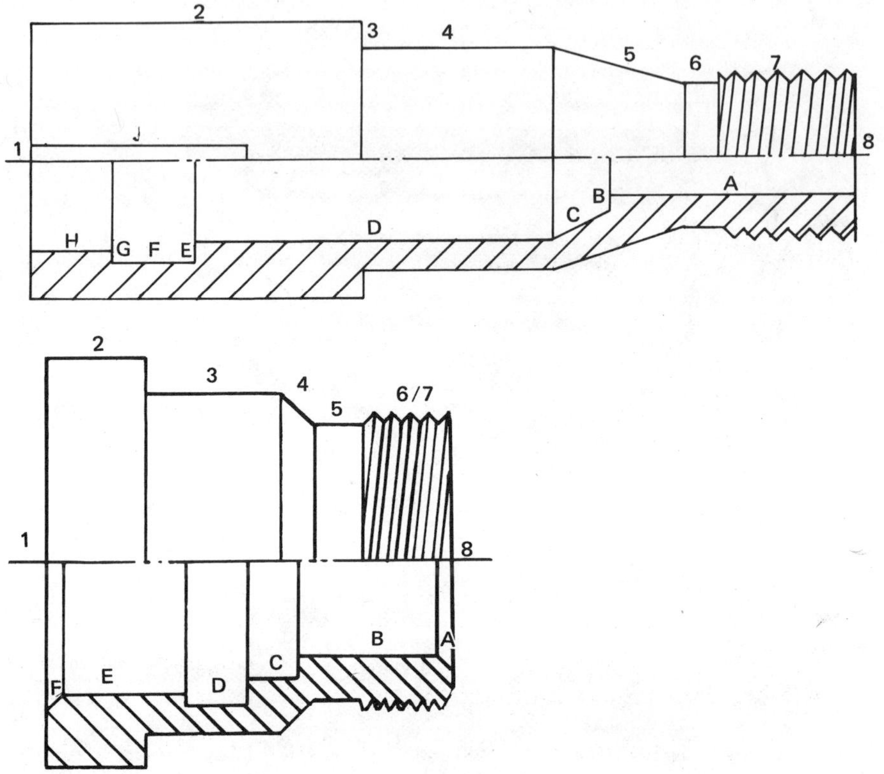

Figure 7.3 Round composite components.

2. Allowed the selection of the following from the used machines found in their own warehouses:
   a. One Herbert capstan lathe capable of accepting a 2-in.-diameter stock through the headstock.
   b. One Herbert capstan lathe to handle other than cutoff bar stock parts.
   c. One No. 2 Cincinnati universal milling machine.
   d. One Cincinnati Carlton 5-ft radial drill.
   e. One two-spindle Avey upright drill press.
   f. One pedestal-type grinder for tool sharpening.

The average age of the equipment was 13 years. Gibs were

tightened and/or replaced. Ways were rescraped where necessary. Headstock and spindle bearings were replaced. Total cost was about $8000.

With the equipment determined, tooling was specified for each tooling station to match the planned processes, developed to create the composite components. Total cost, $10,000.

The specific tooling included the following:

One three-jaw universal chuck capable of maintaining a total of 0.0002 in. total indicator runout (TIR).
One new set of soft jaws for chucking on machined surfaces.
Turret tools with maximum flexibility and accuracy for boring, reaming, threading, and trapaning.
One air-operated vise for the milling machine, as well as air-actuated holding clamps mounted in the T slots when the vise was not to be used.
An index table with four independent jaws for the radial drill, plus an overhung universal drill jig with a range of drill bushings.
A jib crane to handle parts and/or tool and accessories.

The team laid out the equipment and installed the machinery. They then selected three operators and trained them to operate each machine, including the necessary tool setting. This was largely tool adjustment and cutting-tool bit replacement. The operators were also taught how to sharpen drills and single-point tools—both boring bits and turning tools. Cemented carbide tools, reamers, taps, and milling cutters were not to be resharpened in the cell, but sent back to the toolroom at the main shops.

During this training period, a foreman from the main shops participated and was to be the responsible supervisor for the group. Production scheduling was to be done by one person, part time, on a daily basis.

When the cell began operation, it was planned that routing sheets would be prepared for each part. It soon became apparent that this was unnecessary. A manufacturing engineer had developed a standard routing for each composite and the production planner was to delete the unnecessary operations. The time values that had been inserted (as estimated from standard data) were to be added for each order. What actually happened was that the orders were processed in less time than it took to prepare the paperwork.

The production planner stopped the practice and set some average times for each of the composites. He started thinking of the cell as one machine and the hours taken as a group rather than accounting for individual operations.

The foreman spent less and less time in the cell because he really wasn't needed there and was needed in the main shops. It finally shook down to the foreman coming when a problem arose that the three workers in the cell couldn't handle. Ninety percent of the time it was either for a breakdown in equipment or a lack of work.

The majority of parts being processed were for service orders. The normal backlog of service part orders was 12 weeks when processed through the main shops. The reason, of course, is easy to understand in a large batch-order shop running 100 parts or more per order. The workers in the main shops were working on an incentive wage payment plan. The production of one off two or three parts per order cost them lost efficiency, so the production schedules for service parts were always slipping because no one wished to run them. The main shop workers were only too happy to see the small lots out of their lives, and so were the foremen, who were constantly pressured for production efficiency in their departments.

Then problems began to surface. Breakdowns became more and more frequent. The machinery was old and tired before it was made part of the group and, because there was little downtime for setting up, it could not stand the gaff of 90% chip cutting 8 hours per day.

The second problem was that the 12-week backlog of parts had disappeared, and the cell was operating on an hour-to-hour basis. At this time, the two most venerable machines were replaced by "newer" old machines, which had been made available by replacement with newer, high-production machines in the main shops.

The order quantities were increased to 20 parts or less, and with the operators' suggestion for replaceable chip trays, nonferrous parts were added. This served to rebuild the workload to one-shift lead time.

The production cycle from start of work to finished part averaged 3 hours per order. Efficiency of the group based upon frequent performance appraisals was running over 100% on the average. This was in spite of the fact that the operators were on loosely measured day work.

The documents necessary to operate and control the group diminished from 11 pieces of paper to 3. Manufacturing engineering,

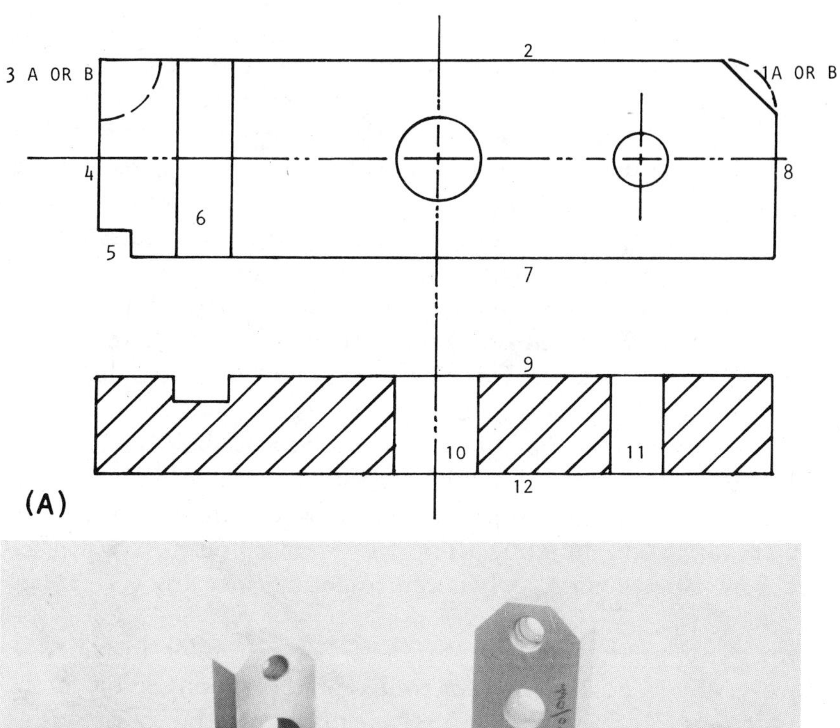

Figure 7.4 Flat composite component and parts made in the same GT module as round parts.

concerned about low utilization of machines (mainly the mill and radial drill), processed another composite—a shape completely different from the original composites. These flat parts (see Figure 7.4) were produced on the mill and the radial, as obviously they are not round. They then created a fourth composite, intending to go to a second-shift operation while accepting larger lot sizes. It was then discovered that many of the new parts of the third round composite required a new balance of workers and machines. A higher machine process capability was also necessary. A decision had to be made whether to enlarge the group of people and add four more machines or to create a second cell. After much persuasion from the original team of consultants and Hopkinsons people, a second cell was created—again, a small group of new machines with four operators and new tooling. A third cell was created eventually—all with benefit, but none to equal the accomplishments of the original cell group working in a nineteenth-century plant building using castoff machinery. A warm and happy glow always comes when remembering that first GT experience.

Imagine—productivity of that first GT cell pilot increased by 240% over the standard times for comparable parts preceding the GT installation. Setup dropped by 85%, as anticipated, but what was not expected was the reduction in machining time, the 100 + % group performance, and the substantial reduction in unit cost. The actual overhead burden rates (GFO) assigned to that first cell were no more than 25% of labor costs, because:

1. Inspection was count verification only, as the operators inspected their own work.
2. Rarely, rarely, was a piece produced out of specification, with small reoperation and/or scrap allowances.

But how does a firm go about investigating and, then, if warranted, installing group technology? In both the examples referred to in this chapter, classification and coding of parts and materials were operational and data visibility assured prior to the installation. There are a variety of ways to get going, but in each instance, data visibility is a prerequisite. This visibility may be acquired by means other than classifying and coding part geometry, but some form of analytical effort akin to classification must be used, if only by association.

The several recognized methods of selecting candidate parts families and/or groupings are:

1. Visual — selection of parts by skilled manufacturing engineers who are intimately acquainted with parts and processes.
2. Analysis of a classified and coded data base of parts which reflects design shape and engineering features.
3. Analysis of a classified and coded data base of parts which reflects method of manufacture.
4. Analysis of unclassified data of parts processes—operation by operation and/or machine.

There are variants of these methods. Instead of purely visual selection from memory, isometric photographs may be taken of parts and machine surfaces and features coded. This is a rudimentary variant of both methods 1 and 2 and is restricted somewhat to a firm that usually makes a number of size/materials variants of similar items.

The most common method employed in the USSR, which has many more installations than those in any other country, is a combination of methods 2 and 3: a hierarchical code (monocode) to simplify and standardize design variety, and a process-oriented code (polycode) that relates how shapes are created and how the engineering features are added to the part. In other works, multiplex codes predominate in the USSR.

Although this was discussed in earlier chapters, it is useful to review the technique from the GT viewpoint. In a total population of 50,000 parts, with a 5000-family code size, there would be only 10 parts average in a family. Obviously, the selection of a single family would not provide a sufficient number of parts to justify a GT cell for low-volume parts only. Thus, for GT, the group level of the code may be used, or perhaps the subclass level, according to prevailing conditions. Specific processing codes may be added to select parts with similar processing opportunities. More often than not, this is not required, and certainly not for the pilot. It is important to keep in mind that how parts are presently processed should not influence the decision as to how they should be made.

Figure 7.5 shows a sample of round parts, taken at random from a large population of piece part items (more than 100,000). The smallest diameter of the parts shown is from ½-in.-diameter stock. The largest

OPERATION TYPE/MACHINE NO.

			Operation	Cut off	Cut off			Turn	Turn	Turn	Turn	Turn	Turn	Turn	Grind	Grind	Mill	Mill	Mill	Drill	Drill	Drill	Bench	Bench	Bench
			Machine No.	691	693	660	800	105	108	152	150	102	151	155	501	510	375	350	351	452	454	455	711	701	732
			Bldg	10-3	10-3	10	?	9-2	9-2	9-2	9-2	9-2	9-2	9-2	9-1	9-2	10-3	9-3	9-2	9-1	?	9-1	9-2	9	8
Drawing No.	Diam.(in)	Length(in)	Material / Description	Power Cutoff Wheel	Band Saw	Babbitt	Heat Treat	19x79 Leonard EL	26X126 Cozo.EL	3 WS HTL	#5T 5L HTL	18X60 Mazer EL	2A WS HTL	5 WS	Landis Univ.	42" Blanchard	3" DeVlieg	A5 Cinn	2/3 Cinn Vert/4KKT	Avery		5-Ft Radial	Burr	Layout	Burr
3-4784D-09001	0.75	4	Inconel	●			●			●					●								●		
5-200FG-01001	3.5	6.125	IR605		●												●						●	●	
5-430AM-15001	4.	0.97	606IT6*(GRANGE)		●						●									●			●	●	
5-034AM-01001	4.5	1.25	IR17		●	●					●							●		●				●	●
3-460GP-01001	4.75	5.125	IR3T4		●							●				●					●		●	●	
5-041JB-02001	5.0	1.125	IR17		●	●							●						●	●			●	●	●
5-506JA-94201	6.0	6.50	IR18		●				●												●			●	
3-459AD-03001	7.5	2.375	IR314		●		●	●								●						●	●		
5-010AA-03001	0.5	10.125	IR176 PIPE XX	●										●					●				●		
3-010DG-03001	0.5	9.063	IR709 PIPE XX	●										●					●				●		

Figure 7.5 Sample analysis of round parts of similar process, as developed from sample process sheets.

diameter is from 7½ in. diameter stock. Lengths range from about 1 in. to a maximum of approximately 10¼ in.

Now, examine the process analysis. Three parts turned on engine lathes could be turned on one turret lathe in a GT cell equipped with the specific tools. In fact, all the parts shown (except 5-200FG-01001 not turned), could be processed on one horizontal turret lathe. It can be seen that turning a 5-in.-diameter part on an engine lathe with a 26-in. swing and a 10½-ft bed length makes little sense when compared with a 7½-in.-diameter part turned on an engine lathe with a 19-in. swing and a 6½-ft bed.

Therefore, to analyze the process flow as currently shown on existing process sheets, and grouping parts by machine type or machine number to select candidate GT parts, simply perpetuates existing, often poor, practices. That most are turned, some are ground, milled, drilled, and all are burred is useful for statistical analysis. It would be necessary to standardize the processes based upon *expected* processing in a GT cell, where engine lathe quantities—usually one or two off—can be justified on a horizontal turret normally devoted to larger quantities because of higher setup costs. Therefore, process analysis in the flow process method is not likely to produce effective end results.

The most practical way to set up a GT pilot cell, and subsequent additional cells as justified, is to use classified and coded data to:

1. Create family composite configurations, as shown in Figure 7.6.
2. Cross-check quantities of parts within the families which are candidates—say, 20 parts or fewer per order.
3. Develop one or more GT composite components from a number of the family composite configurations quantified by total family items within the lot-size restriction established (see Figure 7.7).
4. Process the composite component, select machines, lay out, and install (round parts as in Figure 7.3A and B).
5. Make process picture sheets, as shown in Figure 7.8, identifying geometric surfaces to be machined for operator training.
6. Match specific surfaces on the composite component(s) with tooling stations, specifying the tools, fixtures, and gages (see Figure 7.9).
7. Develop group standard times, using standard data developed for the composite(s) to be run in the cell.
8. Extend times by quantities as forecast or from a prior period's experience.

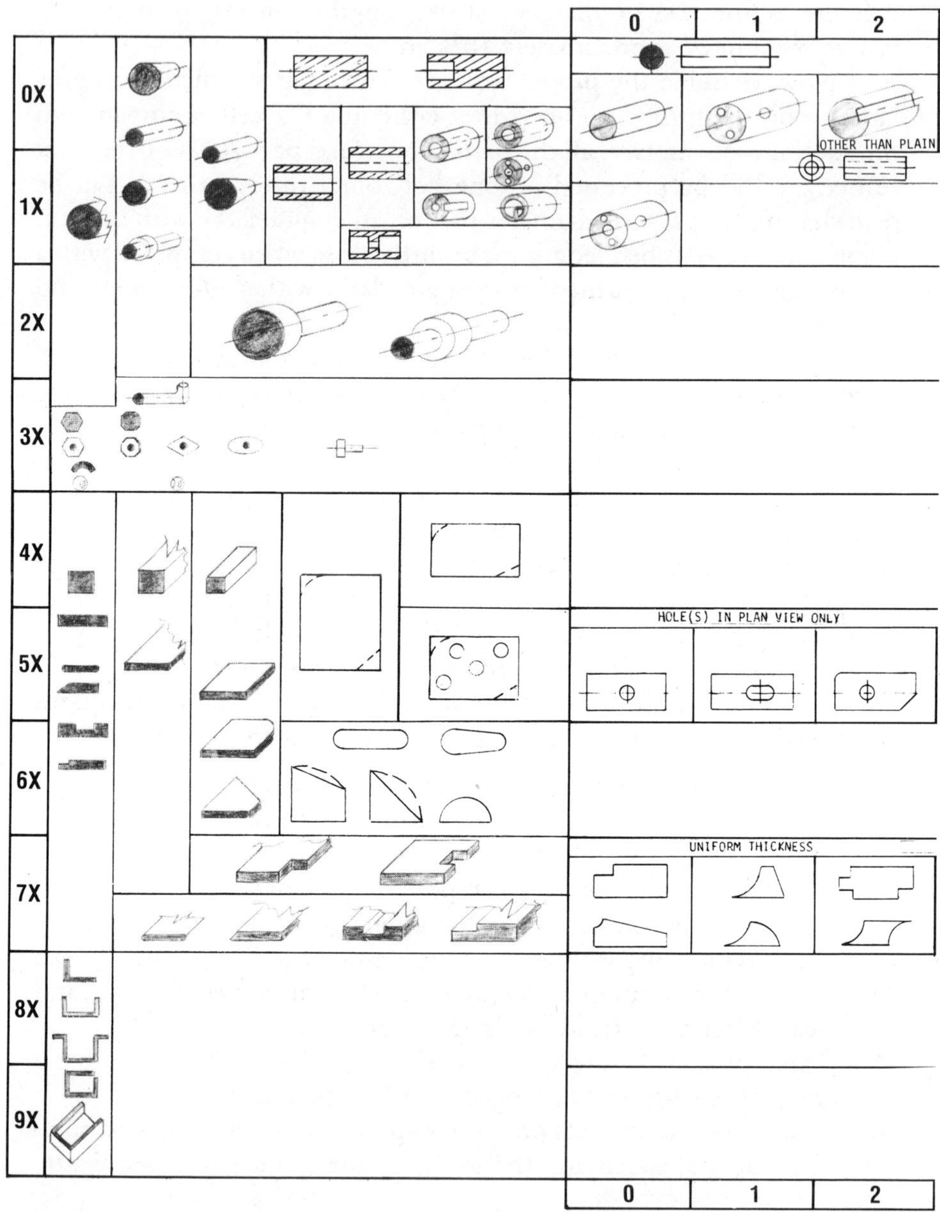

Figure 7.6 Finding family configuration drawings for flat parts. (Copyright © Brisch, Birn & Partners.)

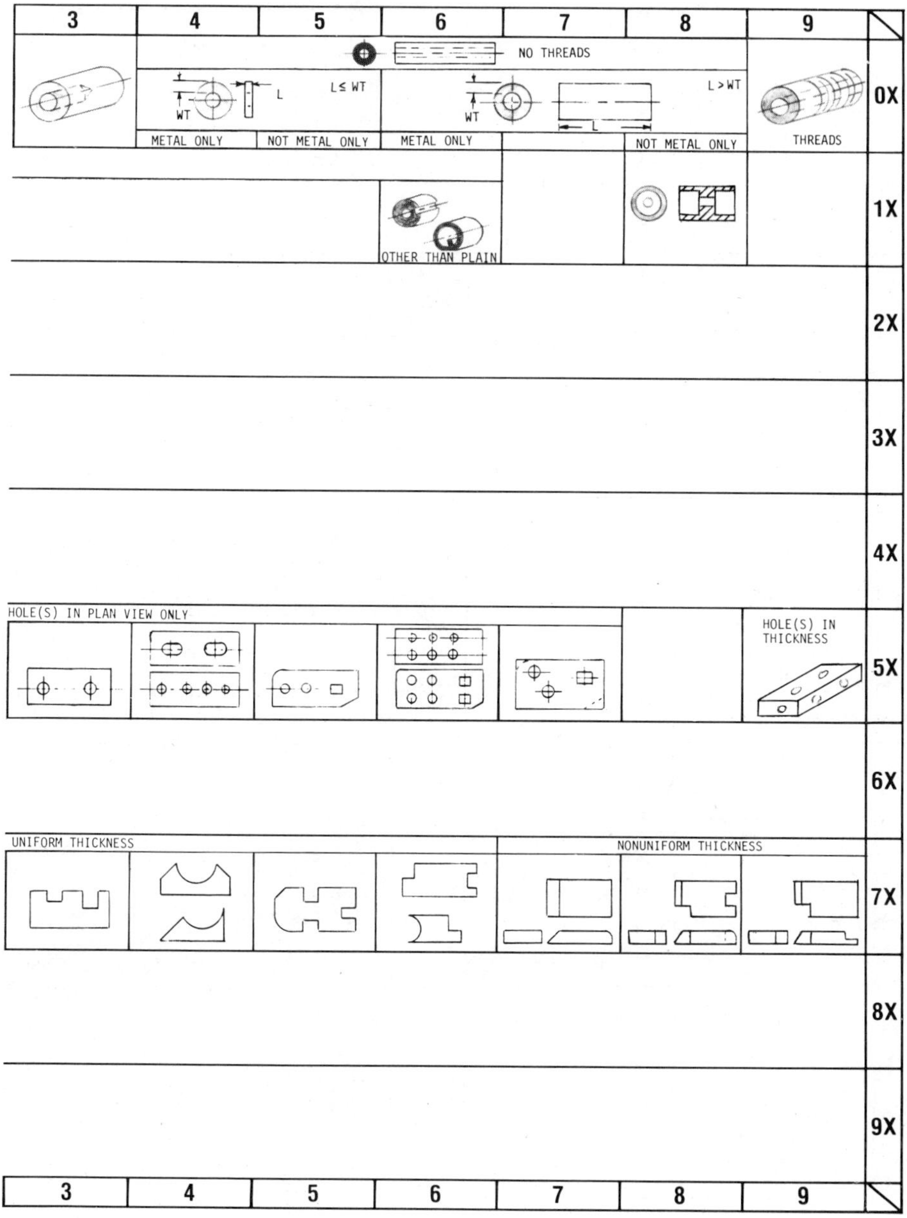

	3	4	5	6	7	8	9
0X		NO THREADS					THREADS
		L≤ WT; WT; L		WT; L > WT; L			
		METAL ONLY	NOT METAL ONLY	METAL ONLY		NOT METAL ONLY	
1X				OTHER THAN PLAIN			
2X							
3X							
4X							
5X	HOLE(S) IN PLAN VIEW ONLY						HOLE(S) IN THICKNESS
6X							
7X	UNIFORM THICKNESS				NONUNIFORM THICKNESS		
8X							
9X							
	3	4	5	6	7	8	9

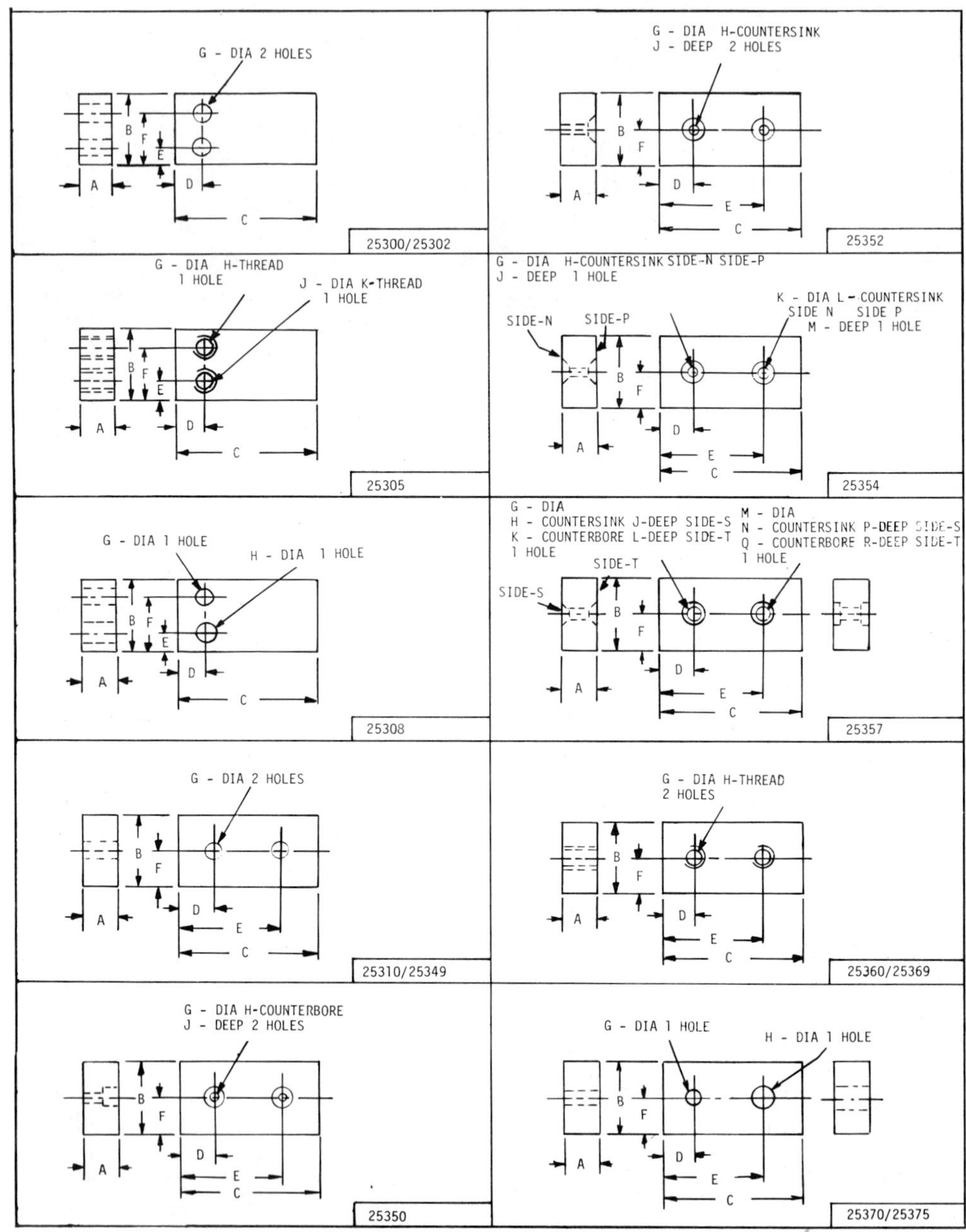

Figure 7.6 (continued)

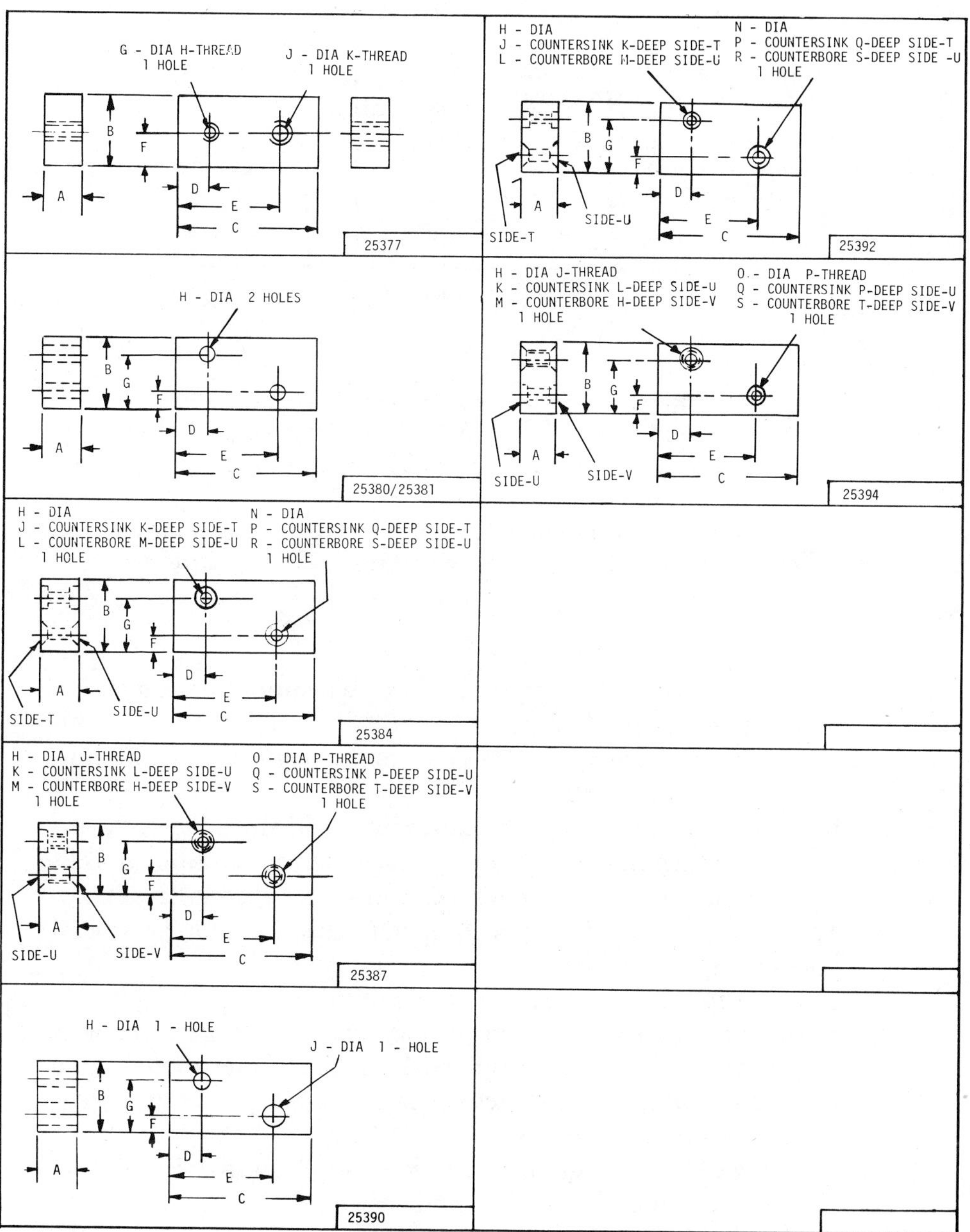
G - DIA H-THREAD
1 HOLE
J - DIA K-THREAD
1 HOLE
B
F
A
D
E
C
25377
H - DIA
J - COUNTERSINK K-DEEP SIDE-T
L - COUNTERBORE M-DEEP SIDE-U
N - DIA
P - COUNTERSINK Q-DEEP SIDE-T
R - COUNTERBORE S-DEEP SIDE -U
1 HOLE
B
G
F
A
D
E
C
SIDE-T
SIDE-U
25392
H - DIA 2 HOLES
B
G
F
A
D
E
C
25380/25381
H - DIA J-THREAD
K - COUNTERSINK L-DEEP SIDE-U
M - COUNTERBORE H-DEEP SIDE-V
1 HOLE
O - DIA P-THREAD
Q - COUNTERSINK P-DEEP SIDE-U
S - COUNTERBORE T-DEEP SIDE-V
1 HOLE
B
G
F
A
D
E
C
SIDE-U
SIDE-V
25394
H - DIA
J - COUNTERSINK K-DEEP SIDE-T
L - COUNTERBORE M-DEEP SIDE-U
1 HOLE
N - DIA
P - COUNTERSINK Q-DEEP SIDE-T
R - COUNTERBORE S-DEEP SIDE-U
1 HOLE
B
G
F
A
D
E
C
SIDE-T
SIDE-U
25384
H - DIA J-THREAD
K - COUNTERSINK L-DEEP SIDE-U
M - COUNTERBORE H-DEEP SIDE-V
1 HOLE
O - DIA P-THREAD
Q - COUNTERSINK P-DEEP SIDE-U
S - COUNTERBORE T-DEEP SIDE-V
1 HOLE
B
G
F
A
D
E
C
SIDE-U
SIDE-V
25387
H - DIA 1 - HOLE
J - DIA 1 - HOLE
B
G
F
A
D
E
C
25390

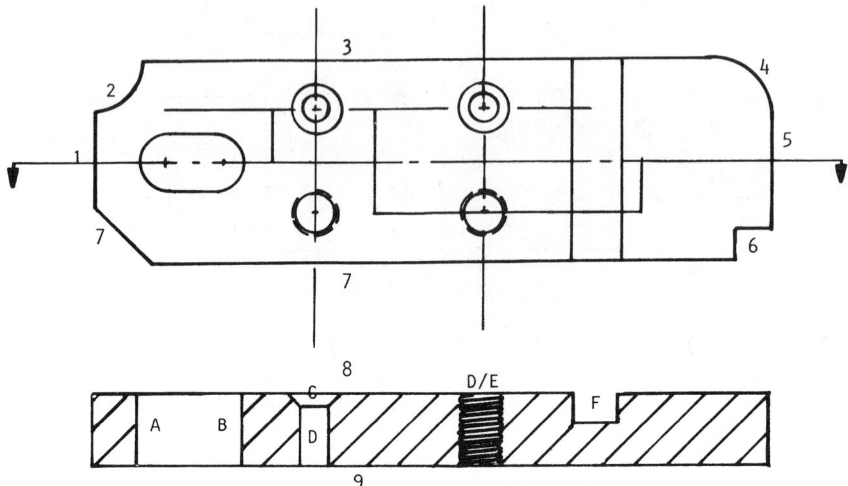

Figure 7.7 Evolved flat composite component: engineering drawing and graphic display. (Graphic display copyright © Brisch, Birn & Partners.)

9. Match hours available to the anticipated schedule of parts by using group times for processing.
10. a. If less than one shift load per month, create additional composite components,
    b. If greater than one shift load, either add shifts, increase cell size (machines and people), establish an additional cell, and/or run excess through the conventional shop layout.
11. Train three or four foremen to operate the GT pilot for several hours a day for 3–4 days.
12. Debug glitches in tooling and/or equipment.
13. Select appropriate number of people—three, four, or as many as are needed—to match anticipated load, *but* never more than a ratio of five machines to three operators.
14. Train them to operate the cell.
15. Load the cell in 2-week units and keep the load inviolate. Allow workers to plan daily schedule and machine assignments.
16. Evaluate the first 2-week unit performance to determine the average time per lot processed by composite component, if more than one.

17. Compare costs per average lot with actual costs when last run through conventional shops, if on a job-order cost system, or measure against standard costs, if on such a cost system.
18. Schedule five more units and repeat steps 11 and 12.
19. Report to management the results of the 12-week trial period with regard to:
    a. Setup cost reduction
    b. Scrap and/or reoperation expense
    c. Each-piece cost comparisons

Figure 7.10 includes most of the steps shown, plus establishment of polycodes for when there is more than one cell operating. This chart, steps 7 through 20, shows the comparable plan on a linear responsibility chart format.

7.4 GT AND JOB DESIGN FOR MOTIVATION

In Chapter 1 reference was made to job design for motivating the new breed of worker. In GT cells, such as at Hopkinsons, this happened quite by accident. They sort of backed into it because no one knew of the Maslow theories at the time of the installation.

When the decision was reached to make the GT pilot installation at Hopkinsons, the reduction in setup time and cost was used as the principal justification. Because it was only a pilot, it was to be an experimental endeavor. Isolated from the main plant as it was kept an experimental air about it from management's viewpoint, but not to the workers assigned to operate the cell.

As the frequency and tenure of supervisory visitation diminished, decisions usually made by supervision were made by the workers. With no setup worker assigned, tool changing, adjustment, and limited setup alterations were made by the workers themselves. They also decided which of them would operate what machine, when, and in what order. In short, they became what we would recognize today as an autonomous group—a group operating as a team with reasonably broad decision-making responsibilities for their own actions.

It is no longer theory that job fragmentation destroys the motivation of workers who must do this nonchallenging work. In Chapter 1 we spoke of the waste of human resources. This was graphically expressed

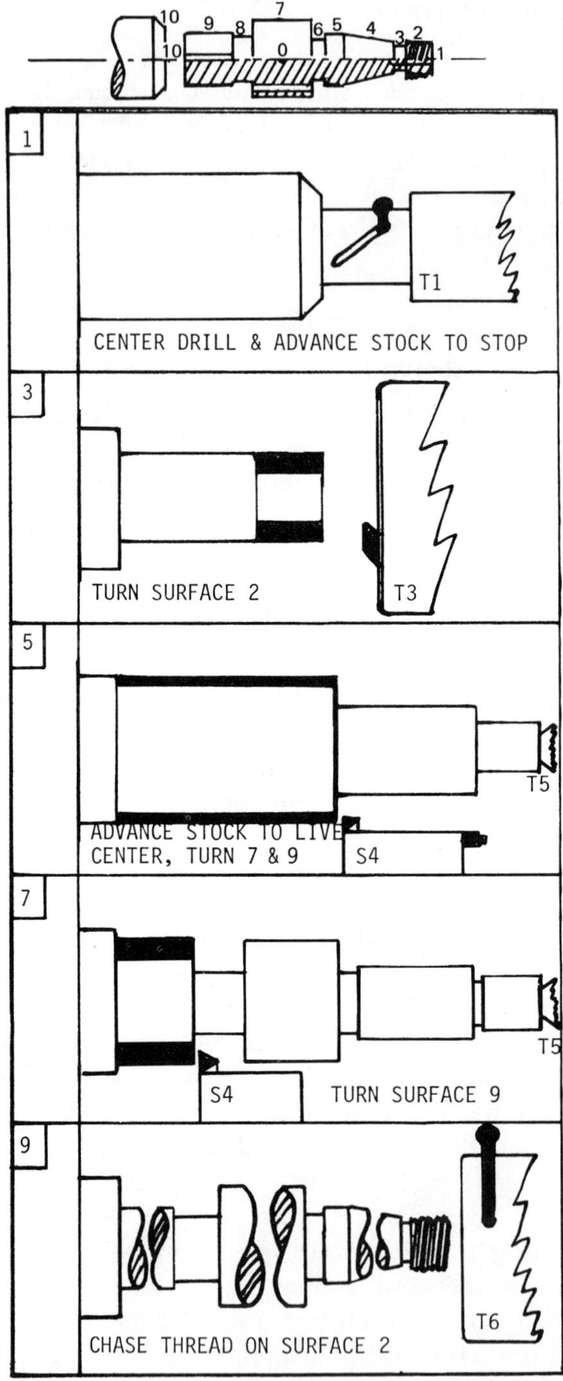

Figure 7.8 Process picture sheet for round composite component shown.

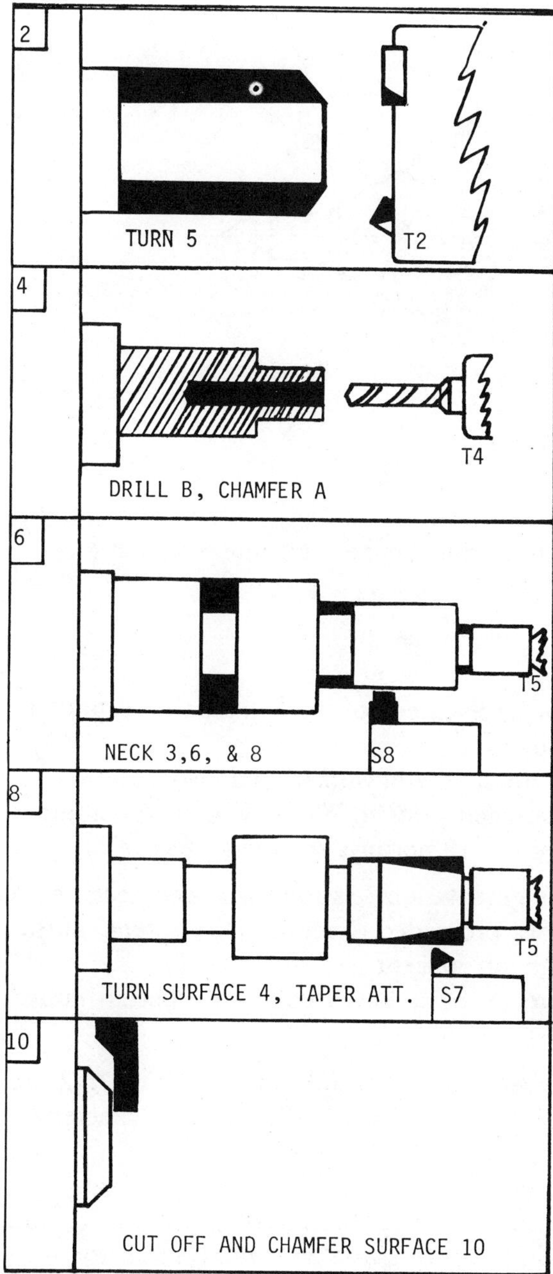

Figure 7.8 (continued)

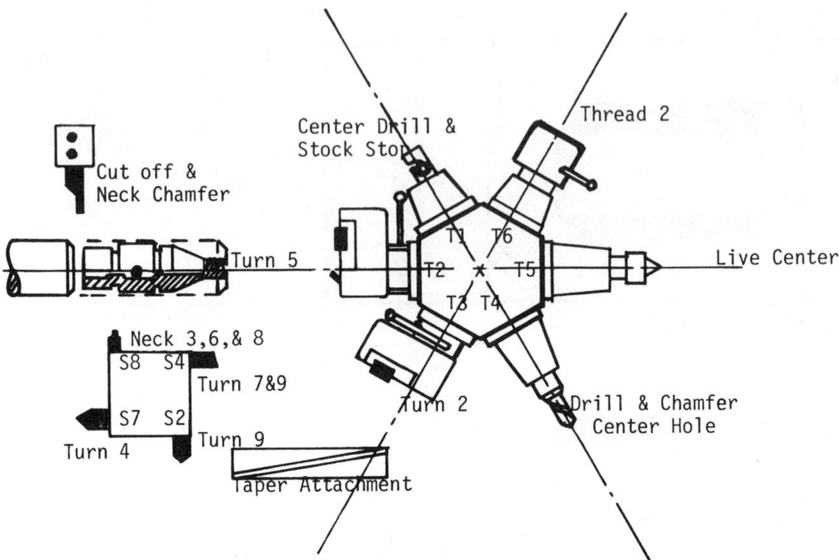

Figure 7.9 Tooling stations to machine round composite component shown.

by Whitsett and relates the evolvement of worker capability prior to World War II to that of today.

Figure 7.11 shows the capabilities of workers to do work and the requirements of jobs as they existed prior to World War II. Capabilities matched requirements. Figure 7.12 portrays the dual effect of:

1. Work simplification (fragmentation), as industrial engineers, with their micro analysis tools, proceeded to reduce job content, pursuing the one motion-one job concept.
2. The increased competence of a better-educated, better-trained work force.

The net result is shown when the two graphs from Figure 7.12 are superimposed in Figure 7.13. The crosshatched segment characterizes the wasted human resource, according to Whitsett.

Admittedly, this is a simplification of a complex problem. Also conceded is the myriad of external forces that exert pressure to restrict productivity. But as stated in Chapter 1, the fact remains that the concept of good industrial engineering held so universally by industrial engineers is, in fact, poor engineering. Although the numbers appear to prove that simplifying work has reduced unit cost of labor input per manufactured product, this has not been the case. Overall, productivity in the United States has not been improving at an ever-increasing rate—it is diminishing.

We have not learned that labor unit cost figures in our plants can indicate a reduction, but variances, absenteeism, labor disturbances characterized by slowdowns, work stoppages, poor quality, grievances, and the like, do not always show up in these figures. They are more likely to appear as variances in general factory overhead.

A. B. Volvo did not make the decision to redesign work and to spend hundreds of millions of kronar to build the Kalmar plant to reduce unit labor costs. They were forced to do something because it was impossible to operate a plant with up to 7000 absentees on a Monday morning. The facts are that labor unit costs are, if anything, higher than in the Gothenburg progressive assembly line. But the improvement in quality and the reduction in rework and scrap more than compensate for the increased unit labor cost effect.

That some pedagogues are claiming that Kalmar is an example of group technology applied to assembly is roughly equivalent to saying that GT is job enrichment. They are both correct, but by accident, not by design. This does not mean that elements common to both techniques are not there. They most certainly are and, if jobs of assembly and machining are designed with both techniques in mind, the best of both can be realized.

It is not our intent to make this a book on job design, but it is important to emphasize the need to create an atmosphere within the GT cell that the workers are a team and that, as a team, they have the freedom to plan how best to use their time and the equipment at their disposal for the most productive results possible. Giving that authority and freedom to innovate and plan has clearly been demonstrated by groups, including GT cell operations, to be fruitful. More of this should be done.

One last bit of advice. The greatest single problem that will have to be resolved concerns acceptance by management of an alien philosphy. It is alien in the sense that most managers have had ingrained in them that work simplification meant job simplification. This is not true, as the

CODE MEANING

1 DOES WORK

2 GENERAL SUPERVISOR

3 DIRECT SUPERVISOR

4 COORDINATE SUPERVISOR

FOR: ELEVATING ORB MOTORS, INC.
YOKAHOMA, MICHIGAN

OF: GROUP TECHNOLOGY PILOT
MANUFACTURING MODULE INSTALLATION

BY: P. E. R. KINS
ON: 12 MARCH 1978

APPROVED
BBP G. M. FORD
ON 13 MARCH 1978
CLIENT OPEL HOLDEN
ON 1 APRIL 1978

SEQ NO.	ACTION DESCRIPTION	PRESIDENT	V.P. PRODUCT DESIGN	V.P. MANUFACTURING	CHIEF MANUFACTURING ENGINEER	CHIEF WORK STANDARDS ENGINEER	VARIETY MANAGER	CHIEF TOOL DESIGNER	PRODUCTION MANAGER	PRODUCTION CONTROLLER	CHIEF QUALITY ASSURRANCE	Q.C. ENGINEER	MANUFACTURING ENGINEER	WORK STUDY ENGINEER	CAD/CAM FUNCTIONAL HEAD	SHOP FOREMAN	PERSONNEL MANAGER	MAINTENANCE MANAGER
		A	B	C	D	E	F	G	H	J	K	L	M	N	P	Q	R	S
1	REVIEW PROCESS IN FAMILY ORDER				3	7	6						1	6	7			
2	SIMPLIFY/STANDARDIZE PROCESSES				3	7	6						1	6	7			
3	PROCESS FAMILY CONFIGURATIONS			7	3	7	6						1	5	7			
4	CREATE MACRO DATA BY FAMILY				3	7	6						6	5	1			
5	CREATE MICRO DATA BY UNIT				3	7	6						6	5	1			
6	POLYCODE FAMILY CONFIGURATIONS				3	7	6						1	6	5			
7	RUN ORDER LOTS, ASCENDING				3	7	7						7	7	1			
8	SELECT G.T. CANDIDATE ITEMS				3	6	1						1	1	6			

5 DECIDES POINT SUBMITTED

6 MUST BE CONSULTED

7 MUST BE NOTIFIED

8 MAY BE CONSULTED

9	CREATE G.T. COMPOSITE COMPONENT		7	2	6	5	1						6	6	7			
10	PROCESS COMP. COMPONENT	7		2	3	7	7	6			5		1	6	6			
11	PREPARE PROCESS PICTURES		7	7	3	7		5			6		1	6	6			
12	SPECIFY MACHINES/EQUIPMENT	5		2	3	7		6	6		6		1	6	7	6		7
13	SPECIFY TOOLING	7		2	3	7		1	7	7	7	7	1	5	7	6		
14	LAYOUT G.T. MODULE	8		7	3	7		5	7	7	7	7	1	6		5		5
15	SET TIME STANDARDS FOR COMPOS.			7	7	3		7	7	7	7	7	6	1	7	5		
16	INSTALL MACHINES/EQUIPMENT			8	8	8							6	6		5		1
17	SET TOOLS ON MACHINES/EQUIP.							3					5	5		1		
18	SELECT OPERATORS												5	5		5	1	
19	TRAIN OPERATORS												6	5		1	3	
20	SET G.T. CYCLE TIMES									3	1	7	7	7		6		
21	SCHED. MOD.IN 2-WEEK UNITS									3	1	7	7	7		6		
22	TEST, TRIAL RUN, DEBUG AS NEC.	7	7	7	3	7	7	7	7	7	7	7	1	5	7	1	7	7
23	RUN FOR 1 MONTH	7	7	7	3	7	7	7	7	7	7	7	7	7	7	6	7	7
24	EVALUATE RESULTS	7	6	6	3	7	6						1	6	7	5		

Figure 7.10 BBP linear responsibility program control chart for installing a GT pilot unit.

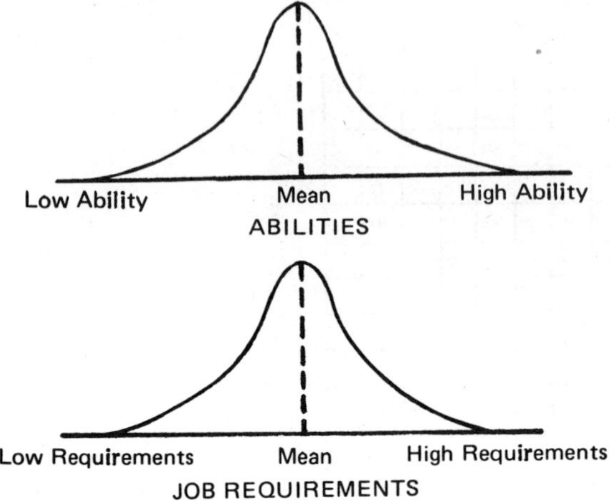

Figure 7.11 Job requirements versus employee skills prior to World War II.

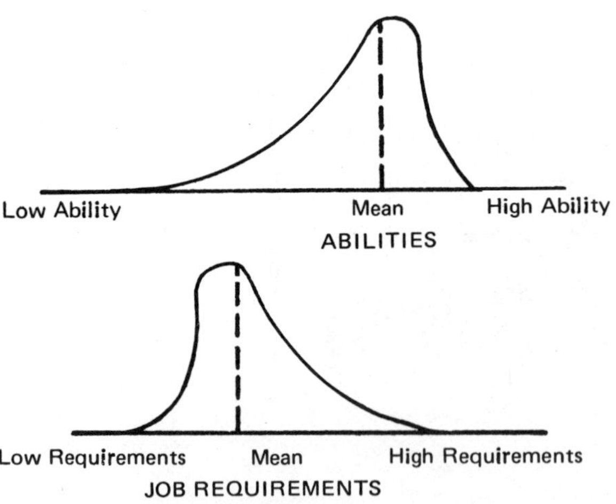

Figure 7.12 Job requirements versus employee skills after World War II.

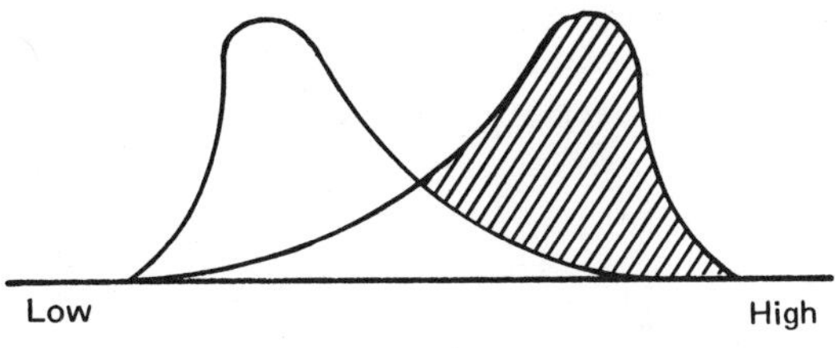

Wasted Human Resources

Figure 7.13 Superimposed curves showing the waste of capabilities of the work force.

two are mutually exclusive. That is, work tasks can be simplified but a job can be enlarged to include additional tasks and responsibilities.

For shop foremen, production planners, and industrial engineers to relinquish some of their traditional responsibilities to the workers is not easy if it is posed as a prerequisite for a successful GT operation. Yet this must happen, but it should happen in a free and unstructured—or at least apparently unstructured—introduction to GT.

Robert N. Ford suggests that discussion-type training meetings be aimed at how the functionaries and managers involved would go about approaching the problem—a sort of brain storming of the subject. With the objectives established as far as what is to be accomplished with GT, a free-wheeling informal meeting or two will set the stage. Then the actual operation of the cell for a few hours by those involved is very useful to make the points in the rationale behind the anatomy of a GT operation.

It is very important that a pilot cell be conducted as an experiment and as a training exercise. If the air of an experiment and participation in the experiment is made to be a real experience and not contrived, experience has shown that management and staff will allow their intellectual curiosity and their loyalties to override their inherent feelings of concern and even mistrust.

That a foreman must assume an advisory role instead of absolute authoritarian is not always easy, but if the situation is kept in an experimental mode, the situation and experience with this new technique

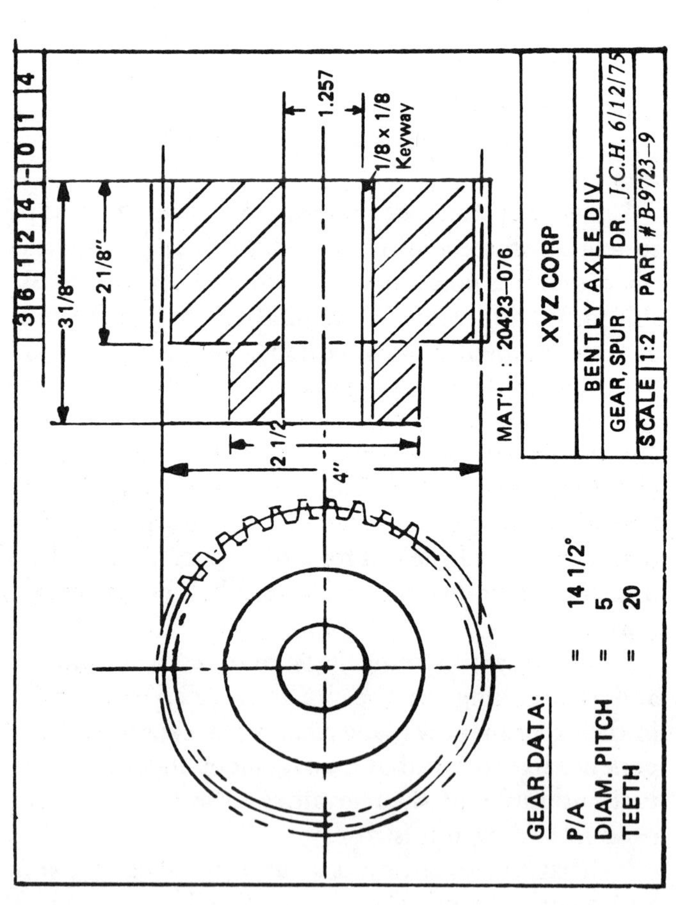
3 6 1 2 4 4 - 0 1 4
3 1/8"
2 1/8"
1.257
1/8 x 1/8
Keyway
2 1/2
4"
MAT'L. : 20423-076
XYZ CORP
BENTLY AXLE DIV.
GEAR, SPUR
DR. J.C.H. 6/12/75
SCALE 1:2
PART # B-9723-9
GEAR DATA:
P/A = 14 1/2°
DIAM. PITCH = 5
TEETH = 20

PART NUMBER / ASSEMBLY ROUTING NUMBER	ACC.	PART OR ASSEMBLY NAME	PAGE OF
36124-014		GEAR, SPUR: HUB ON ONE SIDE	1/1

MATERIAL	RATIO	STOCK DEPT.	DATE OF RECORD	M B	CSTG. WEIGHT
20423-076	7/1		6/26/77		

ENG. CHG.	METHOD CHANGE	PLAN INT.	SUPERSEDES
6/20/77	1		SUPERSEDED BY 8/1/78

	OPER. NO.	MACHINE GROUP	PROD. NO. POS. CODE	EST. PROC. TIME	DESCRIPTION	SET	BY SETTING SET UP Hr	BY SETTING STANDARD Hr
1	030	0001	1022	1.5	PREPARE STOCK AND DELIVER TO 4041		.375	
2	140	4041	0821	3.5	SET 8" ENGINE LATHE: CHANGE CHUCK, SET CUTTING TOOLS, CLEAN UP		.415	
3	177	4041	0821	.8	TURN & FACE HUB-8 RHF/CHMFR: FACE HUB & HUBSIDE GEAR FACE OUT			.635
4	180	4041	0821	.8	END FOR END, CHUCK ON HUB O.D. FACE, TURN O.D., DRILL & REAM I.D.			.422
5	400	1000	0820	1.5	INSPECT & WRITE MOVE ORDER		.225	
6	500	5012	0261	1.7	SET UP HOB: CHANGE ARBOR CALCULATE ADD, ADJUST HOB		.300	
7	510	5072	0261	.4	HOB 20 TEETH, DIAM. PITCH 5, 14½ P.A.			.353
8	515	6095	0797	.5	SET CROWN GRINDER: ADJUST WHEEL, DRESS WHEEL		.150	
9	517	6095	0797	.5	GRIND CROWN			.015
10	401	4047	0797	.5	INSPECT & WRITE MOVE ORDER		.350	
11	700	7770	0031	1.0	SET KEYSEATER		.100	
12	715	7770	0031	.5	CUT KEYWAY			.400
13	902	7000	0031	.5	INSPECT & WRITE MOVE ORDER		1.500	
14								
15								
16				15.7	SHIFTS/LOT		3.415	1.825
17							Hrs/Lot	Hrs/Pc
18								
19								
20					I			

Figure 7.14 Process and production cycle for a typical part to be manufactured: the part and its process sheet to manufacture seven parts per month.

will cause a change in attitude. *Don't rush it!* Make haste slowly and you will benefit in the long term.

7.5 GT AND MATERIALS REQUIREMENT PLANNING

It has been said that GT enables you to make what you want, when you want it, and at the cost you want. The same has also been said for *materials requirement planning* (MRP). To meet assembly requirements in an efficient and timely manner demands that three conditions be satisfied:

1. Dependable, reliable, short-cyle lead times.
2. Quick response to changing needs without increasing costs.
3. The precise and exact quantity of acceptable quality of parts needed for assembly--no more, no less!

As with many an innovative management technique, system, or procedure, the result often does not match the promise. And the reason, just as often, can be shown to be the parochial nature of the undertaking. Perhaps this is why fewer than 10% of all MRP installations produce even 50% of their full potential.

The greatest emphasis has been placed upon the second criterion —fast response. Experts in the field of inventory management have emphasized this as the critical item. Without quick response to reprioritizing, due to any of a variety of reasons, MRP fails to satisfy the third criterion. So a great deal of attention has been directed to the second requirement. Small wonder when the process and production cycle times are examined for a typical part.

The part and its process to run seven parts per month, as shown in Figure 7.14, list 13 operations, of which 9 are machining operations or setup operations. If these are condensed to three work centers and time allowed in each work center is 3 days, a minimum of about 2 weeks is necessary to plan for a finished lot. If a lot is lost, it means a crash plan is necessary to circumvent a delay in assembling on schedule. It is not necessary to elaborate on what that means to plans and schedules. Therefore, the concern is understandable.

The use of GT as a means to contravene or neutralize this problem has proved its worth. However, it cannot be done as a palliative for all poor practices. Why GT is MRP's best ally is its short response time—at most, a shift (see Figure 7.15). This presumes, of course, that the item

PART NUMBER / ASSEMBLY ROUTING NUMBER	ACC.	PART OR ASSEMBLY NAME	PAGE OF
36124-014		GEAR, SPUR: HUB ONE SIDE	1/1

MATERIAL	RATIO	STOCK DEPT.	DATE OF RECORD	M B	CSTG. WEIGHT
20423-076	>20		8/1/78		

ENG. CHG.	METHOD CHANGE	PLAN INT.	SUPERSEDES 6/26/77
	2		SUPERSEDED BY

	OPER. NO.	MACHINE GROUP	PROD. NO. / POS. CODE	EST. PROC. TIME	DESCRIPTION	SET	BY SETTING: SET UP	BY SETTING: STANDARD
1	030	0001	1022	1.0	PREPARE STOCK & DELIVER TO G.T. PRODUCTION UNIT 7		.375	
2	1	P.U.7	4077	1.0	TURN, DRILL, REAM, HOB, CROWN GRIND, KEYSEAT, DEBURR		.600	1.500
3	400	P.U.7	4077	.2	INSPECT & WRITE MOVE ORDER		1.500	
4								
5								
6			X=	2.2	SHIFTS/LOT		2.475	1.500
7							Hrs/Lot	Hrs.Pc
8								
9								
10								
11								
12								
13								
14								
15								
16								
17								
18								
19								
20								

Figure 7.15 Process and production for the same part made in a GT production unit.

to be run falls within the parameters of the established cells' capabilities to produce the item. This is not to beg the point, but to put the positive side forward.

These points are certain:

1. There is less likelihood of lots being lost because there are fewer handlings.
2. The production cycle is 1 day maximum and not 10, 20, or 30 days as under more conventional planning allowances.
3. There is no duplication of setup costs.

Therefore, it can be concluded that a plant with established GT cells can neutralize the problem of schedule disruptions when exceptions to plan occur.

7.6 INSTALLATION TIMETABLE

Earlier, a warning was sounded concerning pilot cell installation time and timing and the need to take each step surely and positively. When Serck Audco Ltd., under the stewardship of Gordon Ranson*, its managing director, installed a total concept of group technology, he wrote the following in a paper published in the United Kingdom:

> Productivity has surged forward and three productivity increases have been awarded across the board to works and staff. In addition, stocks and work in progress relative to sales fell dramatically, and overdue orders are maintained at less than 2 per cent of sales. The function of management at all levels became clear and objective. By far the greater energies were formerly deployed "fire-fighting" amid the morass of outdated information and insular practices. The new operation based on logic, measurement and planning, which stated what was required and when, enabled management to concentrate on the failures of the plan.
>
> This is "management by exception," since concentration was centred on the 5–7 per cent not achieved. Moreover, the 5–7 per cent not achieved became the beginning of the next work load, and was never allowed to be lost and ultimately to be tagged on to some future work load. It must be clear that management, now able to concentrate on their true functions, turned their energies to forward planning and better practices. Once the principles and phi-

*G. R. Ranson, *Production Planning and Control,* Her Majesty's Stationery Office, London, England, 1968.

losophies were established they became self-generating, engendering a forward progressive outlook. One vitally important point which cannot be overemphasized is the truism that the "art of perfect management is the art of perfect delegation." Anyone who fails to delegate and pass responsibility to those appointed to do a job fails miserable in his own task of management. It equally follows that such delegation can only be to people fitted to accept and carry out their function. Any lesser people should, or indeed must, be removed and the appointments properly filled. By releasing the energies of people through responsibility which is rightly theirs, and for which they are paid, you will quickly discover the hidden talents and also the deadwood. No company can be efficient if it has mediocrity at any level and it is the first duty to eliminate it.

We are proud of our achievements to date, but look upon them as the spur for even better practices and improved techniques. The world is our market, and our consistent performance of more than 50 per cent export is a mark of our success. Conversely we buy in world markets, and purchase plant and equipment best suited to our needs, wherever its source.

Our immediate target is to arrange our operation so that raw material moves throughout all its changes of state to the finished stores or despatch of a product, without a pause. We are some way there, but there is some way to go yet.

Finally, let us quickly retrace the steps and the results.

1. We used a team of work study engineers to analyse and review our general operation.
2. We centralised all production planning and used data processing machinery. To permit this we had to use a coded system and establish one drawing and one number per part.
3. We set up a high speed production facility backed by disciplined stores control, with consequent low work in progress and stocks.
4. We reduced our throughput time from an average of twelve weeks to under four weeks, and an overdue customer value less than 2 per cent of sales.
5. We introduced job evaluation and labour mobility together with a share of the benefits of increased productivity to all our employees.
6. We provided wide coverage of induction in all the new techniques, and took great care to explain our plans and carry factory opinion with us.

7. We reduced the number of specials largely by pricing them out.
8. We achieved a new look sales force, based on confidence and cooperation.
9. We achieved a total company outlook based on totally integrated operations.

There is no reason to expect any less if care is taken to plan and execute the installation.

8

Classification and Coding and the Computer

8.0 DATA RETRIEVAL USING THE COMPUTER

The purpose of this chapter is to show various ways in which the computer may be used to retrieve data after they have been classified. As stated in Chapter 3, classification is a scientist's tool that is used to bring like things together by virtue of their similar permanent characteristics (i.e., into families), and then to separate or display them in some manner by their essential differences.

Coding was defined in Chapter 4 as the application of symbols that reflect the classification by its chosen parameters, either singly, or in groups, to identify the classification familes. A serial number may be added to each family member to give it unique identity, if desired. If an item is already identified by another number, the family number will suffice when used in conjunction with the existing identifier.

Information retrieval occurs when the relevant data can be singled out of all the existing data. If the file address of the relevant data is known, the data are immediately accessible and can be made visible. It is when the number is not known and it is desired to find information relating to an item and/or a group of items closely similar that the retrieval problem occurs.

When the computer is used to store information, it demands a precise identity if the information is to be recalled. It should, however, be programmed to provide information by families of related data as well as a unique datum. When interacting with the computer for data (information) retrieval, families of similar (as well as identical) data are preferable to identicals only. It is for this reason that this chapter deals with families of data, so that reasonable choice is offered and specific data not now existing may be derived by interpolation or extrapolation.

There are a number of methods for retrieving data when the file address is not known:

1. By externally classifying the unknown and accessing the computer by its family code.
2. By keyword, in context or out of context (after classification).
3. By interactive classification, with or without codes.

Each of these methods of retrieval will be elaborated upon in some considerable detail in the following sections.

8.1 INTERACTING WITH THE COMPUTER BY CLASSIFICATION CODE NUMBER

The access to family information (i.e., multiple item visibility within a family) can be done quite simply. The specimen datum is classified and coded externally, using a cybernet code key. The family code is entered into the computer using the appropriate program menu code for accessing. A listing of the identifiers representing the items in the family can then be shown in a variety of ways, according to the hardware, peripherals, and needs. The parameters may be few, excluding the obvious in a first-pass analysis, or may show all parameters for selection and/or adding a new item.

It is apparent that this method of accessing on an interactive basis is the least demanding upon the computing equipment, and software costs are lowest. It requires the least machine time when compared to some of the other methods that will be presented. Because the amount of tutorial programming is minimized, the program cost for the software for handling data in this manner is under $10,000 in most cases. In Figure 8.1 you can see the data parameters necessary for an initial review to narrow the search.

In Figure 8.2 other information is provided for the several items that may suffice as chosen from Figure 8.1. This method of folding the

```
CAPACITOR, MICA, SILVER, DIPPED, TYPE CM04                              FAMILY  55301
  RADIAL LEADS MIN LGTH 31.8 MM (1.25 IN) UNLESS INDICATED
  LEAD SPACING 3.6 MM (0.140 IN) LEAD DIA .4 MM (0.0L6 IN)
  TOL + OR - 0.5 PF TO 10 OF 5 PERCENT ABOVE 10 PF                          PAGE  1B
  OPERATING TEMP -55 TO +125 DEGC. TEMP COEF -200 PPM/DEGC
  AVAILABLE RANGE 3 TO 300 PF                                           DATE 10/17/78
  SEE DRAWING 1-157-0000 FOR OTHER PARAMETERS


          APPROVED MAKERS                       OTHER DATA              S SEC  PART NO.  F
NAME           STYLE    NAME           STYLE                            T
                                                                        A
                                                                        T

ARCO ELECTRON  DM 10  CORNELL-DUBILI  CD 10                             P 001 1-161-0651 A
 GENERAL INST  RDM10  SANGAMO ELECT     D10                                              A
ARCO ELECTRON  DM 10  CORNELL-DUBILI  CD 10                             P 003 1-161-0602 A
 GENERAL INST  RDM10  SANGAMO ELECT     D10                                              A
ARCO ELECTRON  DM 10  CORNELL-DUBILI  CD 10                             P 005 1-161-0603 A
 GENERAL INST  RDM10  SANGAMO ELECT     D10                                              A
ARCO ELECTRON  DM 10  CORNELL-DUBILI  CD 10                             P 007 1-161-0604 A
 GENERAL INST  RDM10  SANGAMO ELECT     D10                                              A
```

Figure 8.1 Data parameters to reduce search time and conserve tube space.

```
CAPACITOR, MICA, SILVER, DIPPED, TYPE CM04                                  FAMILY 55301
  RADIAL LEADS MIN LGTH 31.8 MM  L.25 IN  UNLESS INDICATED
  LEAD SPACING 3.6 MM (0.140 IN) LEAD DIA .4 MM (0.016 IN)                     PAGE   1A
  TOL + OR - 0.5 PF TO 10 PF 5 PERCENT ABOVE 10 PF                          DATE 10/17/78
  OPERATING TEMP -55 TO =125 DEGC. TEMP COEF -200 TO +200 PPM/DEGC
  AVAILABLE RANGE 3 TO 300 PF
  SEE DRAWING 1-157-0000 FOR OTHER PARAMETERS

        SECTOR RANGE 000 THRU 999                 ALL

        SEC      ELECTRICAL       CASE DIMENSIONS (MAX)
                   DATA          LENGTH WIDTH THICKNESS
               CAP   WORKING      MM      MM      MM
               (PF)    DC VOLT   (INCH)  (INCH)  (INCH)

        001        3         500   9.1     8.4     4.8
                                  (.360)  (.330)  (.190)
        002        5         500   9.1     8.4     4.8
                                  (.360)  (.330)  (.190)
        003       10         500   9.1     8.4     4.8
                                  (.360)  (.330)  (.190)
        004       15         500   9.1     8.4     4.8
                                  (.360)  (.330)  (.190)
```

Figure 8.2 Folded record of additional information.

total record is dictated by the method of operation (a visual display terminal versus a printer) and the capacity of the terminal. If the visual display terminal is 132 characters wide as opposed to 80 characters, all relevant data can be displayed without folding the record.

The entire array of data can be seen at once on a printout of the data items within a family, as shown in Figure 8.3.

8.2 INTERACTING WITH THE COMPUTER BY KEYWORD (IN OR OUT OF CONTEXT)

It is possible, in some cases, to use keywords (nouns and descriptors) to access the computer for data. Commercial software is available and is used in a number of different ways. In Figure 8.4 the computer has been used to prepare a keyword out of context index for a catalog. The catalog is in hard copy, but obviously could be accessed from the index information if so programmed. Figure 8.4 illustrates the use of a code number, based upon a classification of materials purchased and used in a multicampus university system.

With an adequate tutorial program to assist in the retrieval, it may be practical to use this method, provided that the following conditions pertain:

1. The variety of kinds of items is not too extensive (2000 or fewer different kinds and types).
2. The noun keywords are reasonably standardized.
3. There is adequate familiarity of all parties who will use the system to feel comfortable at an accessing terminal.
4. There is adequate security and control of the file to prevent additions or changes by unauthorized people.

The data shown in Figure 8.5 are *NAND gates.* The information is shown as it appears on a cathode ray tube (CRT) face. The program allows access by either word and/or order.

The electronics industry, through the Joint Electronics Device Engineering Committee, has done an exceptional job in standardizing terminology, performance parameters, and value identification. For this reason, it is a reasonably simple matter to use keywords.

There are, however, some problems even so. When the same device may be used to perform more than one role, it becomes necessary

HYDRAULIC MOTORS — FAMILY 2208

REP 1165 JP AK008 — SECTOR 000

COMMODITIES — PAGE 2

CATALOG LISTING — REV DATE 21/04/78

SECTOR RANGE 000 THRU 599 — AXIAL- SINGLE,DUAL,TRIPLE UNITS

ITEM	PRESSURE RATING BARS	PRESSURE RATING PSI	MOTOR TYPE	INLET PORTS NO	INLET PORTS SIZE	OUTLET PORTS NO	OUTLET PORTS SIZE	DRIVE-SHAFT TYPE	DRIVE-SHAFT SIZE	FLANGE S.A.E. TYPE	MAKER & REFERENCE	STAT	ITEM	USER SDGGDP UARLUD	REFERENCE NUMBER	F/M
045	172.5	2500	SINGLE	1	1.1/4	1	1.	SPLINED	1.1/4	C	HAMWORTHY M2B2110-C5-B1-D	P	045	B	8894016	
009	172.5	2500	SINGLE	1	1.1/4	1	1.1/4	SPLINED	1.1/4	C	HAMWORTHY.M2A 2120-G5 B3-D	X	009		465415	
017	172.5	2500	SINGLE	1	1.1/4	1	1.1/4	SPLINED	1.1/4	C	COMMERCIAL M50A-878-SPBE-OL15-7	P	017	D	8903328	
068	172.5	2500	SINGLE	1	1.1/4	1	1.1/4	SPLINED	1.1/4	C	HAMWORTHY M2B-2120-C5-B1-D	P	068	C	8944812	
001	172.5	2500	SINGLE	1	2.	1	2.	SPLINED	1.1/4	B	HAMWORTHY M2PV2125-C3-B3-D	P	001	D	8900833	
005	172.5	2500	SINGLE	1	2.	1	2.	KEYWAYED	1.1/4	C	HAMWORTHY.M2PV2125-G5-B3-D	P	005	D	551446	
019	172.5	2500	SINGLE	1	2.	1	2.	KEYWAYED	1.1/4	C	HAMWORTHY M2PVS2125-G5-B3-D	N	019	C	570412	
035	172.5	2500	SINGLE	1	2.	1	2.	SPLINED	1.1/4	C	HAMWORTHY M2PV2125-C5-B3-D	P	035	D	8918274	
041	172.5	2500	SINGLE	1	2.	1	2.	SPLINED	1.1/4	C	COMMERCIAL M50B-SPS-898- BI-OR25-7-SP-CAB-10-1	P	041	D	8903863	
043	172.5	2500	SINGLE	1	2.	1	2.	SPLINED	1.1/4	C	HAMWORTHY M2PV2125-C5-B1-D	P	043	BC	8894017 8940923	
057	172.5	2500	SINGLE	1	2.	1	2.	KEYWAYED	1.1/4	C	COMMERCIAL M75A-1878-BE-OX27-11	P	057	C	8943846	
071	207	3000	DUAL	2	1.1/2	2	1.1/2	SPLINED	1.1/4	C	COMMERCIAL M51B-SP-1978 B1 OR 25-78 OR25-1	N	071	A	8594271	
061	207	3000	DUAL	2	1.1/4	2	1.1/4	SPLINED	1.1/4	C	COMMERCIAL M76B 1978 B1-OL25-78-OL25-1	P	061	A	8866217	
031	207	3000	SINGLE	1	1.1/2	1	1.1/2	SPLINED	1.1/4	B	HAMWORTHY MPV2216-C3-B1-D	P	031	D	8906246	
063	207	3000	SINGLE	1	1.1/4	1	1.1/4	SPLINED	1.1/4	C	COMMERCIAL M76A 1978 B1-OL30-7	P	063	A	8866218	

 FAMILY 2208-000

Figure 8.3 Catalog hard copy of 132-character record. (System copyright © 1976 Brisch, Birn & Partners.)

UNIVERSITY OF CALIFORNIA CATALOG FOR LOS ANGELES

Entry	Code
BALLS, Cotton/Rayon Balls, Incl. Round Sponges	57574
Toilet Tank Fittings Including Balls, Floats, Guides, Levers, Lift Wires, Overflow Tubes, And Risers	27711
BALSAM Bottles, Balsam Incl Immersion Oil	44039
BAND Saw Blades, Metal Cutting, Machine Operated	35261
BANKERS Pins, Straight. Bankers And Dressmakers	20110
BARREL/MORTICE Bolts, Barrel/Mortice	22331
BASIN, Faucets, Basin, Lavatory And Sink	27273
P-Traps, Sink Or Basin, Brass, Chrome Plated	29571
Wrenches, Basin, Crescent, Hex And Pipe, Adjustable	36650
BASINS, Emesis And Wash, Incl. Tubs, Foot, Plastic	57120
BASKET, Liners Can And Basket, Cloth, Plastic And Treated Paper	63859
BASKETS, Test Tube	46011
BATHROOM, Cleaners, Tile, Porcelain And Bathroom, Incl. Bowl Cleaners	63501
BATHS, Sitz Baths, Disposable,	57126
BATTERIES, Carbon-Zinc, Primary, Non-Rechargeable	30011
Flashlights, 2-Cell, Incl. Penlight, W/O Batteries,	32325
BATTERY, Clips, Test & Battery, Eyelet, Flag And Tag, Screw Connected	33359
BATTS, Cotton, Rolls And Batts, Incl. Coiler And Plugging	57517
BEADS, Glass	44801
BEAKERS, W/Graduations & Spout, Griffin Low Form,	44027
BED LINEN SAVERS Bed Pads And Bed Linen Savers	65261
BED PADS And Bed Linen Savers	65261
BED PANS, Incl. Covers And Drapes	57130
BEDSIDE, Bags, Patient Bedside, Sanitary & Waste Disposal	57111
BEES Waxes And Paraffin, Incl. Bees Wax, Tissue Prep And Embedding, Paraffin	12050
BELLS, Call. Desk And Counter, Tap Type. Manually Operated	77110
BLADES, Knives And Replacement Blades, Carton, Edger, Linoleum And X-Acto Type, Including Knife Sets	36315
Surgical, Non-Sterile, Carbon Steel	56423
Surgical, Sterile, Carbon & Stainless Steel, Incl. Dermatome	56421
Surgical, Sterile, Disposable, Stainless Steel, Incl. Dermatome	56422
BLADES, TONGUE, Non-Sterile,	54400
BLEACH, Dry And Liquid	63533
BLOOD, Administration Sets & Accessories, For Blood, Solutions And Urology, Incl. Armboards, Sleeves, Tourniquets, Extension Tubes, Hemocoils, Murphy Drip Tubes, And I.V. Bottles And Caps	57411
Lancets, Blood, Sterile	56443
BLOTTER COVERS	72275
BLOTTING Pads	72273
Paper	72271
BOARDS Paper Cutters / Trimming Boards	72435
BOLTS, Barrel/Mortice	22331
Carriage, Square Neck, Round Head, With Nut, Unc Thread Steel, Commercial Low Carbon, Plain Finish	20810
Eye, Open Stock Not Welded. No Shoulder, Unc Thread, Plated Steel, With Nuts	20841
Machine, Hexagon Head, With Nut, Unc Thread Steel Commercial Low Carbon, Plain Finish	20800
Sheeting, Plastic, In Rolls Or Bolts, Incl. Acetate, Polyethylene, Polyurethane, Vinyl, And Parafilm	17452
Stove, Round Head, Slotted, With Nuts, Unf Thread Steel, Commercial Low Carbon, Zinc Plate	21226
Stove, Slotted Head, With Nut, Ns Thread Steel, Commercial Low Carbon, Zinc Plated	21526
Stove, Slotted Head, With Nut, Unc Thread Steel, Commercial Low Carbon Zinc Plated	20816
Surface And Pull-Chain	22314
Toggle, Gravity Type Wing, Unc Threaded Steel, Zinc Plated	22021
Toggle, Spring-Wing Type, Unc Threaded Steel, Zinc Plated	22020

Figure 8.4 Keyword out-of-context index.

INTEGRATED CIRCUITS
INCLUDING FUNCTIONS OF
GATING,COUNTING, ENCODING/
DECODING, SHIFT REGISTERING
FLIP FLOPPING

SECTOR RANGE 001/199
GATES

TYPE	INPUT	STATUS	PROP DELAY MAX NS T LH	TP HL	I SUP-PLY NA TYP/ MAX ICC	LOAD-ING FAN IN/ OUT	I IN-PUT MAX ■MA■ IIL	V OUT TYP VOH/ VOL	SECTOR	PART NUMBER/ MAKER NAME MAKER NUMBER
NAND,TTL	SINGL8	P	10	12	6.5 10.0	1.25 12.5	-2.0	3.5 0.2	107	1-590-1029/ TEX INS SN7430J FAIRCHI 7430 MOTOROL MC7430L
NAND,TTL	SINGL8	P	22	15	3.0 6.0	1.0 10.0	-1.6	3.4 0.2	099	1-590-1021/ TEX INS SN74S30J SIGNETJ N74S30F
NAND,TTL	TRIP 3	N	10	10	19.5 30.0	1.25 12.5	-2.0	3.5 0.2	007	1-590-1031/ TEX INS SN74H10 FAIRCHI 74H10DC
NAND,TTL	QUAD2	P	20	20	2.4 4.4	0.23 5.00	-0.36	3.4 0.3	021	1-590-1543/ TEX INS SN74LS00J FAIRCHI 74LS00DC

Figure 8.5 NAND gate data retrieved by keyword in/out of context. Shown are integrated circuits, including functions of gating, counting, encoding/decoding, shift registering, and flip-flopping.

to identify these multiple missions and add these to the descriptors for keyword retrieval.

The major problem for anyone organizing data bases is the true identity of what something is and what it does. It is a difficult and tedious job unless there is reference to an obtaining standard and/or excellent documentation created within the organization using the items in question.

Figure 8.6 is taken from a classified data base of Figure 8.4, the keyword out-of-context index used by a state university system. To use

```
C:  READY-ENTER PERSONAL CODE FOR MENU SELECTION
U:  1013HY
C:  READY-P SELECTOR FOR ONE SYSTEM IN LIST
    1.BERKI
    2.CALCODE
    3.DROBS
    4.UNUS
U:  2.
C:  READY-P SELECTOR FOR ENTRY MODE
    1.CALCODE FAMILY NO.
    2.CAMPUS I.D. PART NO.
    3.MAKER/VENDOR AND NO.
    4.NOUN KEYWORD AND DESCRIPTORS
U:  4.
C:  READY-ENTER KEYWORD AND TWO DESCRIPTORS MAX.
U:  PANS, BED
C:  BED PANS, BED PAN COVERS/DRAPES ARE IN CALCODE FAMILY 57130-5 ITEMS AS LISTED
    1.ENAMEL, VITREOUS, BLUE/GREEN OUTER, WHITE INTERIOR          SECTOR 107
    2.STAINLESS STEEL                                             SECTOR 101
    3.PLASTIC, CRYOLAC                                            SECTOR 111
    4.COVERS, PLASTIC                                             SECTOR 521
    5.DRAPES, REP PLASTICIZED                                     SECTOR 537
```

Figure 8.6 Computer keyword retrieval of classified and coded data.

the computer to retrieve by keyword would require extensive and expensive software because of the tutorial instructions and commands to enable the user to isolate the items wanted.

In Figure 8.6 it is evident that if one were trying to retrieve data on *bedpans,* this would pose no problem. In Figure 8.4 are items shown with the word *bolt* in their descriptions. Because this is a keyword out-of-context library system and printed as an alphabetic index, it can be seen that quite diverse items are included. To retrieve a bolt that is a fastener, say *bolt, carriage, square neck, round head, UNC thread, commercial grade, low carbon, plain finish* (neck length has been disregarded, as it is displayed in the file information for the line item where it applies) from all the items with bolt in their title or description would require extensive tutorial support, either in the computer itself or from hard-copy instruction. A typical problem to contend with can be illustrated with "Bolts, Carriage." There are two ANSI standards, B18.5-1952, Reaffirmed 1959, which show the style and physical dimensions. Material and finish are not shown. Neither is the thread series. In one standard, the heading reads *ROUND HEAD SHORT SQUARE NECK BOLTS.* In the other, the heading reads *ROUND HEAD SQUARE NECK BOLTS (Square Neck Carriage Bolts).* There is also reference to undersize body bolts.

There are also standards for bolts with similar configurations and/or alternatives for similar application with names as follows: The reference in Figure 8.4 to *BOLTS, BARREL/MORTICE* is for dead bolts in a door—not fasteners in the usual sense.

Again, let it be stressed that none of these problems is insurmountable, but the cost effectiveness is certainly suspect when compared with alternative means to retrieve with the computer and/or to use indices in hard copy, film, or stored magnetically.

The alphabetic index shown in Figure 8.7 lists data references by keyword in an indent style reflecting the parameter order of the classification. It may be stored magnetically in this fashion. It may also be produced in hard copy, or on film, as an operating instruction tutor.

The alphabetic index of Figure 8.8, truncated in description to 24 characters, is produced by computer but is not—repeat, not—a mutually exclusive means to isolate a specific family. Families 67800, 67802, and 67804 have identical descriptions. If code capacity were not a factor (as it is in the firm using the classification code number as a part number), the entire family could be listed in a single family. More than

	FAMILY NUMBER
UNIONS	
Alloy Steel Pipe Fittings	2564
Carbon Steel Pipe Fittings	2544
Iron Pipe Fittings	2524
Rotary	2618
Swivel Joint	2618
UNIVERSAL JOINTS	2385
VACUUM GAUGES	
Measuring Vacuum Only	2928
Measuring Vacuum or Pressure	2928
VACUUM PUMPS	2400
VALVES ACTUATED BY A CHANGE IN DIRECTION	2453
VALVES ACTUATED BY A CHANGE IN FLOW	2455
VALVES ACTUATED BY A CHANGE IN LEVEL	2452
VALVES ACTUATED BY A CHANGE IN PRESSURE	2454
VALVES	
Angle	2425
Ball	2420
Butterfly	2426
Check	2453
Diaphragm	2429
Directional Control	
Two Ways	2430

Figure 8.7 Keyword indent method by computer-generated microfiche. (Copyright © Brisch, Birn & Partners.)

2000 individual items are involved in the several families. It would means (1) lengthening the identification code to cater for the largest family, or (2) dividing the families by size split, or (3) expanding the noun descriptor as allowable. Each firm's circumstances and needs vary. What the best solution is can be determined only by study of the problems. Then the options can be compared for convenience versus cost and the best solution recommended.

8.3 INTERACTIVE CLASSIFICATION

To program a computer to interact to classify a part was done in a 1966 joint project with a computer manufacturer. The capability to do this, therefore, not in dispute. What may be questioned is cost effectiveness.

ALPHABETIC DESCRIPTION SEQUENCE

Code No.	Description
73143	CAM FOLLOWER, BALL TYPE
73140	CAM FOLLOWER, NEEDLE TYPE
73146	CAM FOLLOWER, NEEDLE TYPE
92374	CAMERA/PROJECTOR ACCESS
92330	CAMERA, POLAROID TYPE
73584	CAMS
67820	CAPACITOR, AIR TUNING
67600	CAPACITOR, ALUM
67620	CAPACITOR, ALUM, WIRE LEAD
67800	CAPACITOR, CERAMIC TRIM
67802	CAPACITOR, CERAMIC TRIM
67804	CAPACITOR, CERAMIC TRIM
67700	CAPACITOR, CERAMIC TUBUL
67710	CAPACITOR, DISC, TO 50 WV
67712	CAPACITOR, DISC, TO 100 WV
67714	CAPACITOR, DISC TO 500 WV
67715	CAPACITOR, DISC OVER 500
67716	CAPACITOR, DISC, 1000 WV
67717	CAPACITOR, DISC, OVER 1000
67760	CAPACITOR, DUAL ELEC
67720	CAPACITOR, FEEDTHROUGH
67740	CAPACITOR, MICA
67737	CAPACITOR, OIL FILLED
67729	CAPACITOR, OTHER THAN 677X
67732	CAPACITOR, PAPER
67730	CAPACITOR, PAPER, TUBULAR
67750	CAPACITOR, PLASTIC & FILM
67625	CAPACITOR, PLUG-IN ALUM
67660	CAPACITOR, TANT, DISC
67640	CAPACITOR, TANT, TO 10 WV
67641	CAPACITOR, TANT, TO 15 WV
67642	CAPACITOR, TANT,TO 20 WV
67643	CAPACITOR, TANT, TO 25 WV
67644	CAPACITOR, TANT, TO 35 WV
67645	CAPACITOR, TANT,TO 100 WV
67646	CAPACITOR, TANT, TO 100 WV
67647	CAPACITOR, TANT, OVER 100
67880	CAPACITOR, VAR. GLASS VAC.
69090	CARBON ARC LAMPS
91106	CARDS,TAGS,ADH LABELS
85770	CASE,LINEAR ADJ

Figure 8.8 Truncated keyword index, created by computer.

In any population of items to be captured, the language used (i.e., the signs, symbols, abbreviations, terms, and the like) requires precise identification. Even so, the complexity of the parameters of the items to be classified have direct bearing on the cost and complexity of the program to do it. A university in Utah has succeeded in doing this for general application of shape-classified single piece parts and the materials from which they are manufactured. It is a good piece of work as far as it goes. There are, however, some problems which have not been addressed. Parts classified by function are, as of this writing, not dealt with, and it is doubtful that it will be feasible to do this, because of the wide variations from user to user.

It is not uncommon to find more of the classification devoted to function of the part than to shape of the part. G. A. B. Edwards, collaborating with A. T. Fathelden, of Manchester University, Manchester, England, have corroborated this in their paper on work piece statistics research.

For classification of any kind to be useful requires that each decision to be taken be mutually exclusive and binary (yes, it is; or no, it is not). This was demonstrated in Chapter 3, where it was also pointed out that the population to be classified must be embraced totally within the bounds defined for the population. Also stressed was the need to choose classification parameters that are permanent, unchanging, and readily discernible.

The attribute block exercise was designed to illustrate logical classification. The emphasis was on the logic chain rather than on computerized retrieval. The same exercise can be used to illustrate how the computer can be programmed to accept the classification logic and its binary, mutually exclusive decisions.

Figure 8.9 shows the logic tree (classification steps) for the simple attribute block exercise. The total population is embraced. The decisions are binary and mutually exclusive. They are based upon readily discernible attributes (parameters) and variables (values) per attribute. The classification logic represents a single point of view. Therefore, all conditions are satisfied for a sound classification using the bipolar notation.

To find any unique item will require no less than four decisions and no more than eight decisions with a pure chance average of six. Expressed mathematically, the number of decisions to be made in this simplistic problem is as follows:

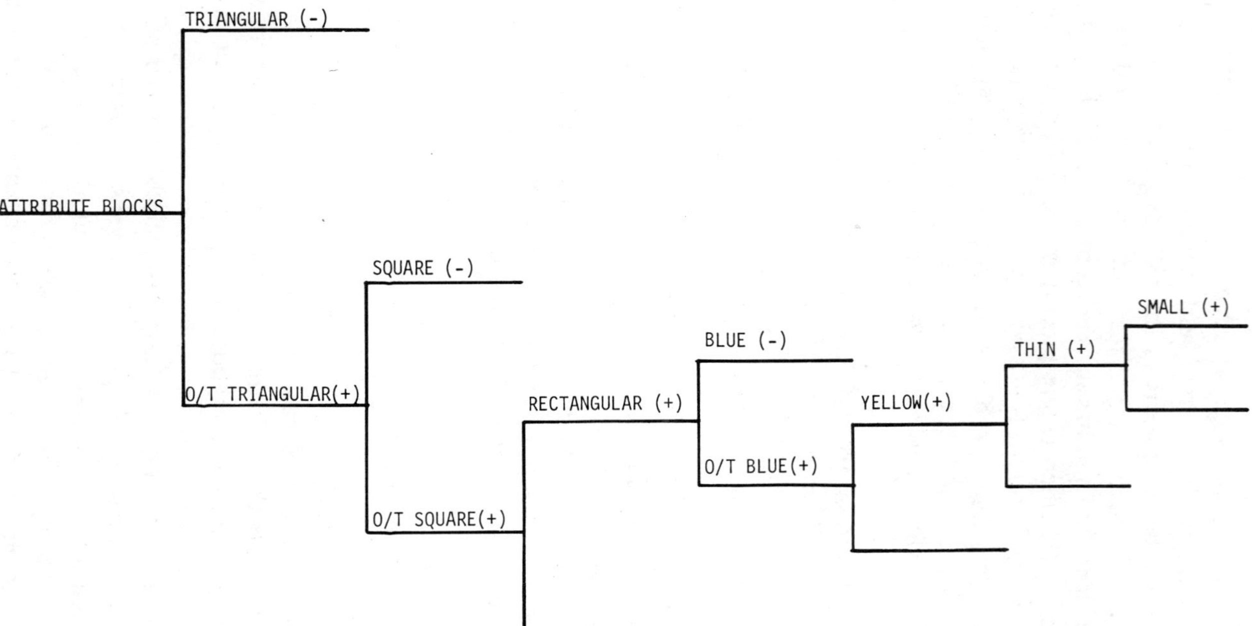

Figure 8.9 Bipolar logic tree for computer-assisted classification and coding.

$$\text{Decisions (maximum)} = \sum_{i=1}^{N} (S_i - 1)$$

where

N = number of attributes
$S_i \triangleq$ number of states (variables) an attribute may have

$$\text{Decisions (minimum)} = \sum^{N}$$

$$\text{Decisions (average)} = \frac{\sum_{i=1}^{N} (S_i - 1) + \sum^{N}}{2}$$

To program this problem may require as little as 4 hr of writing and test to no more than 8–10 hr, according to the language in which it is written.

The time to access and retrieve the datum desired, using the computer interactively to search the population of blocks, is on the order of 2 min. Remember, though, that it is a classification of known quantities. If a search were made of the same shape, size, and thickness, but of a fourth color, the population does not include it except as other than red, yellow, or blue.

For this reason, among others, it is recommended to use a different form of tree to classify a diverse and perhaps infinite universe of items. In the preceding examples, the tree form was purely binary in structure: that is, "yes, it is" ends the search of variables in the attribute question.

The inquiry is made variable by variable. When a "yes" match up is made, the program chains to the next attribute and its variables. The binary polar tree allows only two states per branch.

Another and much more useful form is the n-polar tree. This is the tree form used in monocoding, as developed by Gombinski and Brisch, in the late 1940s. Whereas the bipolar tree is acceptable for a fixed-field classification such as the Linnaean taxonomy, the n-polar tree is more practical for the flexible-field classification method. It is not difficult to distinguish between the two and it can be seen that while the decisions may be the same, the program is simpler and easy to develop.

In the pure binary polar tree, only two states are visible. That is, for each attribute each variable is shown in sequence until a match up is made to progress to the next attribute. We express the decision tree for attribute shape as a series of questions (e.g., is the shape round?). If the answer is "no," the next variable in sequence stored behind the shape attribute is produced. This is a fixed and rigid classification structure.

In the n-polar tree, all the variables per attribute are shown simultaneously. This is the form used by Gombinski in his monocoding procedure. When Brisch originally hewed to the Linnaean line, with some minor variations, Gombinski preferred the flexible form. He did so because it is more useful for coding and code-length conservation.

The n-polar tree has a decided advantage in that:

1. All variables per attribute are seen simulataneously.
2. The scope of the classification is immediately known.
3. Fewer fields are required when coding.
4. Retrieval time and computer time per transaction are minimized.

Figure 8.10 shows the display that would be seen for the shape attribute in the block display using the n-polar tree. The programming for the n-polar tree form, with built in tutorials, requires a bit more time than the bipolar tree for the attributes, largely because of the tutorials needed to assist the inquirer.

Professionals in classification and coding have mixed views on the subject of interactive classification (and coding). The need to do this is largely to overcome resistance to imposed systems. This is a personal observation after discussions with client firms.

One firm in the aerospace industry felt that resistance by design engineers to accept the new system of variety control made interactive classification mandatory. Not so, says a 20-year user of classification and coding in the construction machinery industry. Wayne Smith, recently retired after many years with Caterpillar Tractor Company, stated the following in a recent interview:

> I have consistently given my support to the use of the computer in systems designed to improve engineering and manufacturing operations. One cardinal rule which has always guided me is—that system is best which is realistically cost effective. Too often, a theoretical solution to a systems problem bogs down because of the human element.
>
> As any machine designer [and systems designer] knows, it takes twice the design time to debug the working model. Systems

```
C:  READY-P SELECTOR-ONE FROM LIST
    1.TRIANGULAR
    2.SQUARE
    3.RECTANGULAR
    4.HEXAGONAL
    5.CYLINDRICAL/ROUND

U:  3.

C:  READY-P SELECTOR-ONE FROM LIST
    1.BLUE
    2.YELLOW
    3.RED

U:  2.

C:  READY-P SELECTOR-ONE FROM LIST
    1.THIN
    2.THICK

U:  1

C:  READY-P SELECTOR-ONE FROM LIST
    1.SMALL
    2.LARGE

U:  1.

C:  3211
```

Figure 8.10 N-polar logic tree transaction record from the CRT face to computer assist classification and coding.

people tend to overlook the analogy when it comes to people oriented systems. In the design of machinery, the problems are mechanical, or electrical, or hydraulic, subject to finite physical laws.

Systems work problems are in the realm of the infinite because the problems are largely people problems. And, while there may be psychological guidelines to handle individuals and groups, the greatest ingredients are time, participation, patience, and persistence.

We have never had a coding quality problem at Caterpillar. However, it took three years to encourage the organization to use the system to anything like its full potential. To classify interactively from remote terminals may have some hidden psychological benefits. From where I sit, after twenty years of experience with classification and coding, I think it will still require patience, time,and pervasive, persistent persuasion and participation to achieve the desired results.

8.4 YOU PAYS YOUR MONEY AND YOU TAKES YOUR CHOICE

The root of the question is not whether a computer is to be used in data retrieval, but how. That the computer and classified and coded data are compatible, even synergistic used together, is not disputed. It would appear that to use the computer to classify and code may have benefit as a training aid. In fact, programmed instruction by computer, with graphics capabilities as part of the program, is unsurpassed as a training aid. However, once the knowledge is acquired and the skills are sharpened, there is undoubtedly the economics of diminishing returns to contend with. It is at this point that one must judge the long-range benefit.

Essentially, the problem resolves itself to the principle of the division of labor, or a variant thereof. The Adam Smith-Fayol-Taylor concept of the division of labor is very much disputed by behavioral scientists. In designing today's jobs, the mentality and education of contemporary workers mitigate the need to separate the work of the hands from the work of the head, as prescribed by Taylor et al. However, it appears not to be specialist versus generalist as much as what is the best method of classifying and coding to use the computer. So all methods should be investigated.

Some methods not mentioned previously include the use of closed circuit television and voice communication; another includes the use of sensing devices, scanners, and the like, for configuration and/or alphanumeric data recognition; and another links micrography, graphics, and the computer for merged microfiche production of graphic and alphanumeric data.

On balance, how you choose to use the computer to retrieve will depend upon many factors, a few of which are:

The dispersion of the inquirers
The organizational spread of the inquirers

The kind of data being retrieved (graphical and/or alphameric data)
Equipment availability
The magnitude of the data base

Once these variables can be put into balance, the system can be designed that is best suited to the particular need. Anything less than this is to adopt the technology solution and then fit the problem to it—just the reverse of the proper solution.

9

CAD/CAM and Classification and Coding

9.0 CAD/CAM: WHEN IT ALL BEGAN

Most writers track activities that have led to the origination of an innovative solution to a management problem. All too often, the evolutionary chain of events that finally sprang an elightened solution to a long vexing problem was not even known by the problem-solving innovator. Be that as it may, it is most revealing to trace the events that underlie the emerging technology to improve our productivity which is known as computer-aided design and computer-aided manufacturing (CAD/CAM).

In Chapter 5, the role of Eli Whitney in the standardization of designed componentry and parts interchangeability was discussed. Whitney is probably best remembered for inventing the cotton gin. However, the standardization of parts for his U.S. Army musket contract was an even more important event. That contract was the first known to include language that related the tenets of mass production. It took Whitney far longer to complete the contract than planned (8 years instead of 2). However, the accomplishment is no less amazing and was a quantum step forward. The enlightened creativity of Whitney coupled with the Watt steam engine to free industry from the rivers' edges, gave birth to the Industrial Revolution.

Samuel Colt and his partner Elisha Root capitalized on Whitney's concepts. Their contribution was the establishment of the forerunner, or prototype, of the great American machine shop. By the year 1855, their

Hardford shops boasted more than 1400 machine tools of various genre. These tools were equipped with holding fixtures and jigs designed to hold specific shapes of parts.

These tooling refinements distinguished the Colt works from most of the others. It enabled production rates and output of finished arms that was unique for the time. This large outpouring is all the more amazing when considered with the fact that the machine tool industry was in its infancy. As is the usual case, need begets solution to fulfill the need. So, when two of Colt's employees, Amos Whitney and Francis Pratt, saw the need for a new kind of machine, they quit Colt and set up their own firm to make the machine. As most everyone is aware, both the Colt and Pratt and Whitney firms are still in business.

By 1876, the first automatic lathe had been developed. The machine advanced and retracted tools by a mechanically programmed device that its inventor, C. M. Spencer, called a "brain wheel."

Shortly thereafter, Frank L. Cone and Herbert S. Wilson, both of the Windsor Machine Company, designed a turret lathe and a hand screw machine. George L. Gridley, who had designed an automatic screw machine (single spindle), joined them. By 1910, they expanded the automatic screw machine part of the business to include the first multispindle machine, using the Geneva wheel to index the spindles. The Spencer and Gridley innovations were truly the first automated machine tools. They were programmed (set up) to operate more or less unattended, except for occasional stock replenishment, inspecting the work, and adjusting the tools, as required. Meanwhile, in 1884, in Cincinnati, the forerunner of Cincinnati Milacron, Inc., was born.

It was machinery of this kind that made early twentieth-century mass production of automobiles, farm equipment, and other basic high-ticket products accessible for mass-market consumption.

It only follows, then, that for machine tool technology to advance productivity, engineers would have to concentrate on the one weak point inherent in all machinery of any degree of sophistication. Specifically, the elements of setup time and cost to change over to a new part had to be addressed. This is especially necessary today, as we have departed from the mass production of a single product to a variety of model production. Examine the lot sizes of a typical machinery works, as shown in Figure 9.1.

The latter stages of World War II saw the development of equipment that could perform extraordinary feats in metal removal. Heavier, higher-horsepower-driven machines using cemented carbide tools

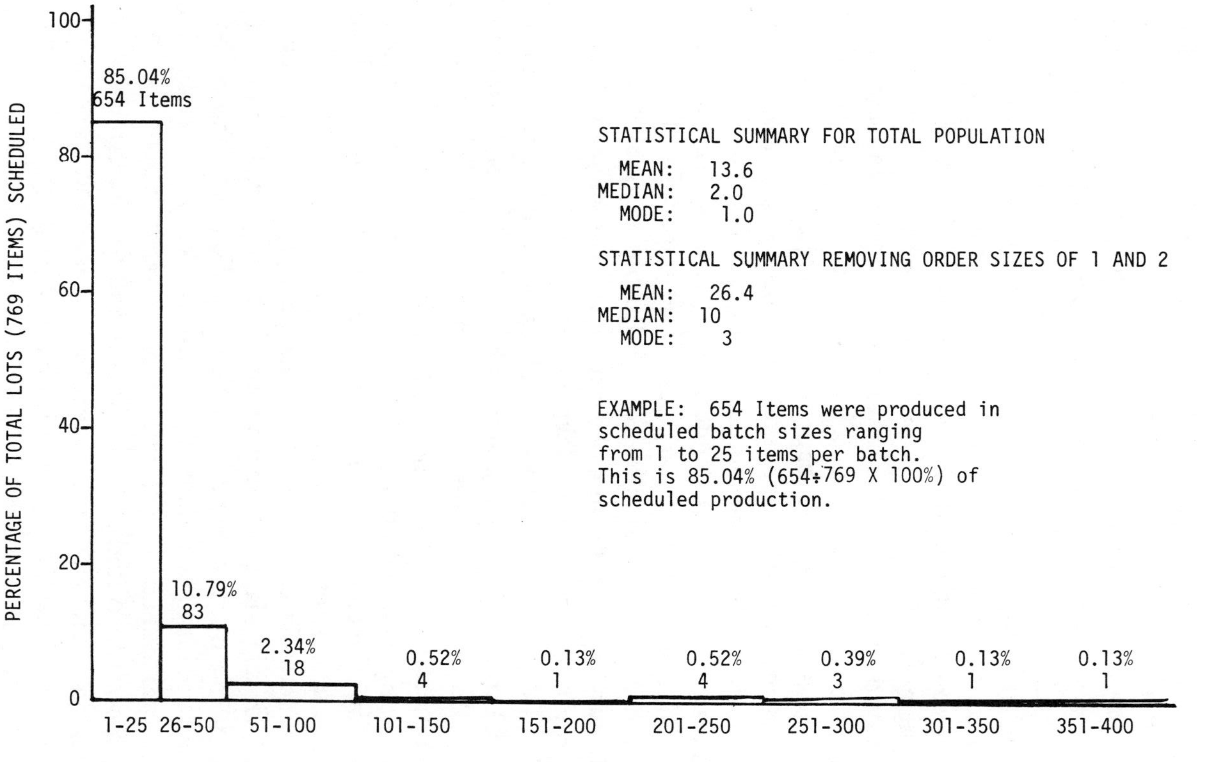

Figure 9.1 Scheduled batch size as a percentage of total production lots (769). (Study based on 5/10/78 shop order quantity report.)

removed metal at a prodigious rate. This allowed designers to design parts using formerly difficult to machine alloys and to cut hardened materials. With minor exceptions, however, machinery was just bigger and better versions of fundamental machine types—great for cutting in straight lines and at angles, but virtually impossible (with a few exceptions such as template lathes and millers) to machine a smooth curve. Designers therefore avoided designing parts that required such surfaces.

J. T. Parsons, in the period between the end of World War II and the beginning of the Korean war, postulated that a method to machine a smooth curved surface could be developed by other than handmade mechanical patterns controlling tracer lathes and millers. Parsons' theory was based upon the premise that any point on a curve could be represented by mathematical expressions of the X-Y coordinates of each of the points. By punching the coordinates for each point on the curve into a card readable by a translating device (control), the machine could move the work from point to point, creating a contoured surface.

The U.S. Air Force was impressed with Parsons' theories and let a contract to Massachusetts Institute of Technology to develop a system of numerical controls employing the Parsons' concept.

M.I.T., in their Servo Mechanisms laboratory, converted a Cincinnati Hydrotel milling machine (tracer-type miller) and built the first two-axis, single-plane, point-to-point numerically controlled machine. The milling head was always on point and the table holding the work moved, adjusting axially and transversely to produce the desired surface. Figure 9.2 shows the points of reference, the movement, and the milling head location in a point-to-point system.

The numerically controlled (NC) machine responds to an input medium which is in a language that a control device can read, interpret, register, and, in turn, command a machine to follow. The earliest NC machines were point to point and came on the market in late 1956. The punch card was displaced by punched paper tape. It soon became necessary, however, to find a more efficient and foolproof method of writing programs. The awkwardness of the language and methods for programming the machines were making it difficult to utilize NC machines to anywhere near their capacity and their capabilities.

M.I.T. set out to develop a symbolic language that would simplify the geometric and process data expression. The result was APT, an acronym for Automatic Programmed Tools. This open-ended approach has been improved upon and updated without making obsolete programs written earlier.

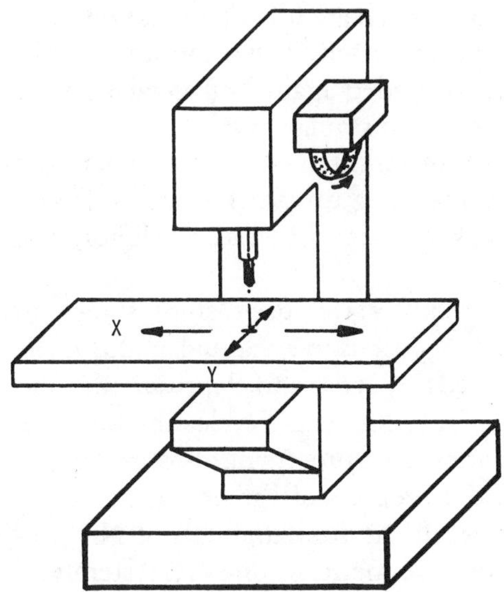

Figure 9.2 Point-to-point numerically controlled machine tool.

As improvements to language and programming software were made, so were new and improved control devices designed. New design concepts for machine tools were developed to accept the enlarged capabilities. The new technology soon covered about every type of machine tool that was traditional to shape materials.

With the improvement in software came new NC systems. Hardwired controls and the machines they control also improved. The multimission machine tool that combines a variety of machining operations, such as turning, drilling, milling, boring, tapping, and counterboring, in a single machine, on as many as five axes, resulted.

In the early part of the 1970s, makers of machine tools were beginning to think in broader concepts. The analog devices inherent in most control systems were too restrictive. Digital computers had been used to control production in process-oriented industries. But the role of the computer in controlling simple, straightforward, on-off, metering, and controlling temperature-pressure and salinity is hardly a test of its use in the machining environment.

Contouring parts, for example, involves at least three axis controls. These are the most complex programs to write. A system using a computer to relate geometry and a tool path to create it is a true test not only of software ingenuity, but of the computer itself. The computer permits storage of the data for recall when it is needed again, thus obviating the laborious and costly programming step for similar parts.

In contouring, the interpolation of infinitely tiny increments from one point in space to another is demanding on programmer/software and the machine control. Therefore, the computer serves a valuable role to ease that problem.

The control of any machine requires the basic steps shown in schematic flow chart form in Figure 9.3. Just as in a manual analogy, the operator sets the tools; selects the feeds, speeds, and depth of cut; and then checks, correcting if necessary. In the machine cycle, the machine monitors both the geometric formation and processing information. It continually feeds back (or should) the progress of the operation.

This book is not intended to teach the fundamentals of NC programming or to hold forth on the relative merits of one NC system versus another. It is the intent of this chapter, however, to show the chicken-egg relationship of machining/designing and its influence on product design. Because a machinist can now make contoured parts as easily as straight turned or milled parts, so can the designer create drawings for contoured parts. Minimum-section, ideal part solutions using contours obviates the need to design multiple parts to do the same thing, reliance upon the founding operation to create them, or the costly process of hand finishing such parts.

As Computer Numerical Control (CNC) and Direct Numerical Control (DNC) systems evolve, so does the need to computer-aid the design aspect. For the most part, these related problems suffered the usual time juxtaposition. Just as in aircraft development, the airplane designer has always been dependent upon the power plant. As a result, aircraft are designed to utilize an engine rather than the other way around.

As difficult-to-express surfaces are created, the designer must communicate in a means compatible to the NC programmer as to what is wanted. The computer is capable of doing this with ease, but the problem and solution were not approached in this manner. Thus, computer-aided design, which was first employed by the Boeing Company to design the rivet patterns on the B-47 bomber more than 30 years ago, did not progress in step with computer-aided manufacturing.

The first commercial systems that were claimed to be computer-aided design were for designing wiring harnesses, wiring diagrams, tube

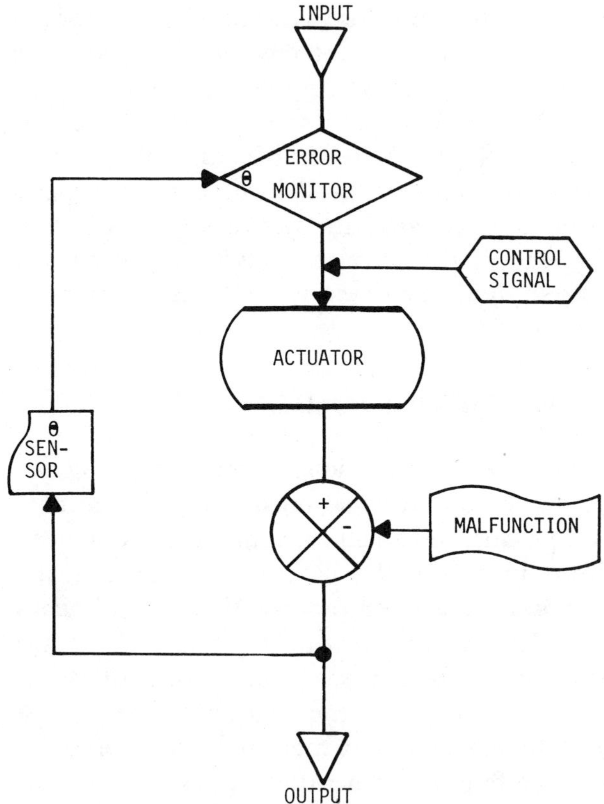

Figure 9.3 Closed-loop NC control system.

bending, hydraulic schematics, and the like. Architectural and civil engineering programs have also been available for 10 or more years. Software, hardware, and data storage costs, however, were inhibiting factors. The minicomputer, a small-scale machine that could be dedicated to a single problem solution, sparked a flurry of activity in the development of CAD for parts and assembly design, for two reasons:

1. Low hardware cost enabled more equipment to be available to experimenters.
2. Resulting developments increased the market for systems based upon minicomputers. This has especially benefited computer-assisted parts and assembly design and manufacturing applications.

New languages and modifications to main frames and peripherals were necessary to use the new computers, but for CAD/CAM great progress resulted. A number of new languages have emerged for geometric models and three-dimensional designing of complex part forms. Computer graphics technology has been adapted to make use of the new languages, equipment, and software, with built-in tutorials to make training and use of these. Minisystems are available in the marketplace in strength. It is interesting to note that all the systems developed up to now require some capability for retrieving selective data (e.g., geometric configurations and/or assemblies). Retrieval is an inherent problem common to all, but solved by few.

9.1 CAD/CAM: ONE PROBLEM, ONE SOLUTION

CAD and CAM have been viewed for too long as separate problems/solutions. This is due primarily to traditional organizational treatment of the two functions (i.e., product design and manufacturing engineering). The diagram shown in Figure 9.4 considers the design and manufacture of a product as an integrated process. While traditionally we have separated the problems associated with the design of product from those related to its manufacture, it was a forced solution rather than an optimum solution. In most firms, the relations between those who create the design and those who have to make it range from peaceful coexistence to outright warfare—or so it would appear.

When a part is created, it is possible to create some of the special tooling necessary to make the part at the same time, and create the data to operate the NC/CNC/DNC equipment to make the part. For example, a formed part design with the assistance of the computer will show the tool path to create the cavity of the dies to produce it and the tapes to make it happen. The system, therefore, creates not only the part drawing but, in addition, the NC data form necessary to make the tools that will, in turn, make the part. This confirms rather dramatically that there may be two problems of our own making, but that one solution will suffice.

9.2 HOW A CLASSIFIED AND CODED DATA BASE IMPROVES CAD/CAM PRODUCTIVITY

The chief executive officer of a multinational corporation was asked to comment after a CAD/CAM system was demonstrated:

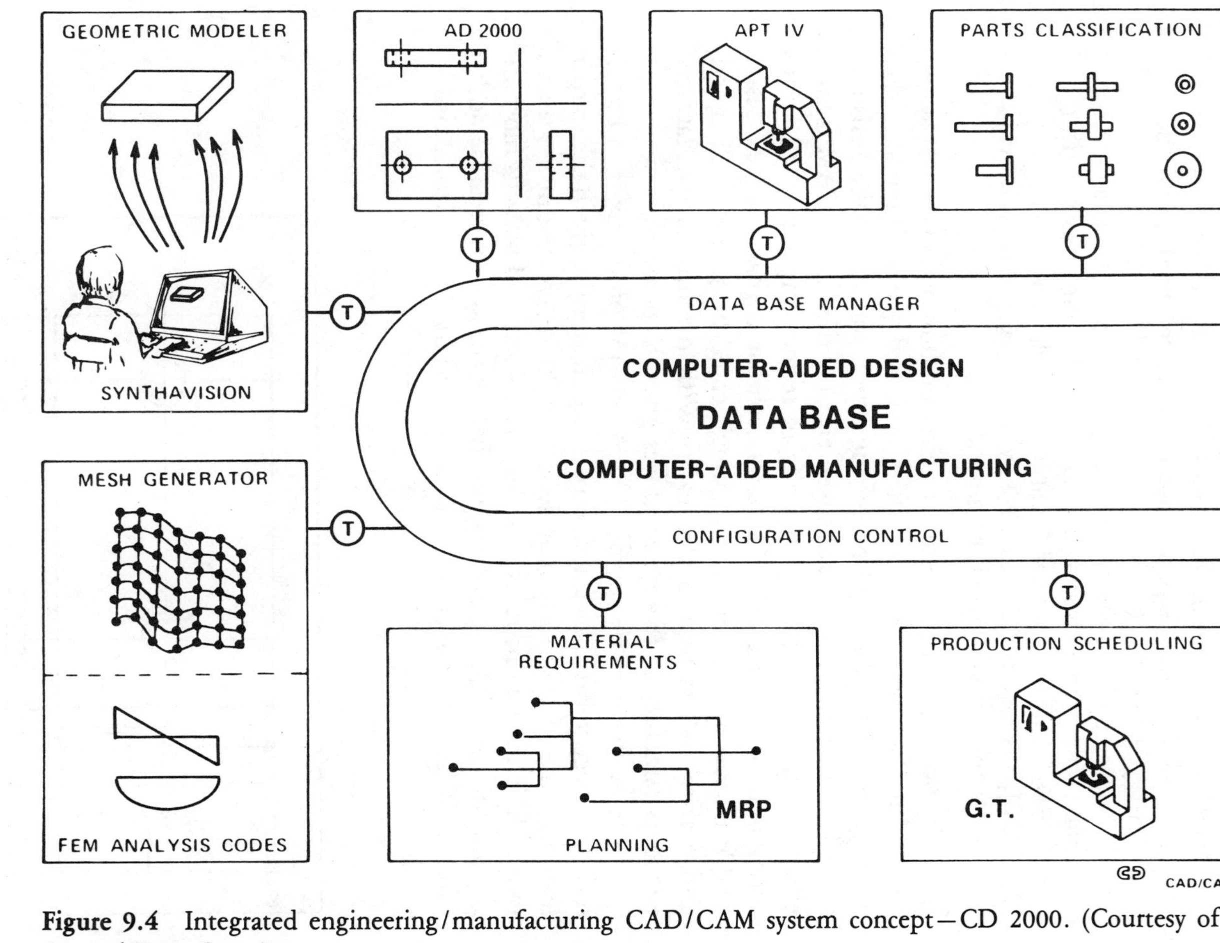

Figure 9.4 Integrated engineering/manufacturing CAD/CAM system concept—CD 2000. (Courtesy of Control Data Corp.).

> Impressive! Most impressive! . . . [CAD/CAM] takes the tedious hand work out of the designer's job—the drafting. Gives us more of his brain time—that's good. Should cut our new model [introduction] cycle as well.
>
> One problem that I can't see an answer to is: How does this affect our product standardization policy? I can't connect this [CAD/CAM] with our parts control standards. If we can't control this [parts, assemblies, materials, variety], [our] worldwide standards policy won't get off the ground.

The operations vice president described how a CAD/CAM system would interface with the simplified preferred parts alphameric data already stored in the central processing unit—and retrievable. Also included was the planned enhancement of the CAD/CAM system efficiency through storage graphically of refined models called family configurations (see Figure 9.5). It was estimated that complex design problem areas, such as gears and springs, could be analyzed, designed, and drafted in less than 6 min each. With the link to standardized elements established, the approval was given to proceed.

The accuracy of the predicted CAD/CAM productivity enhancement has been verified by actual experience. It is still early in the startup learning curve, but already gear designs are being created in less than 7 min each. This demonstrates how design data retrieval plays its most important role in a CAD/CAM system. Before the system creates another variant, the specific dimension, materials, and the like are made visible.

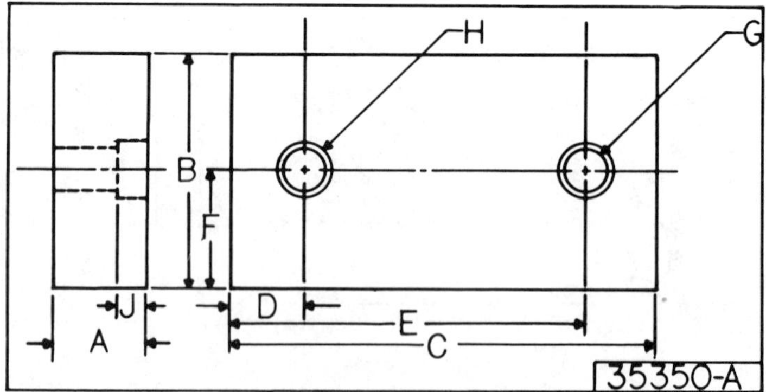

Figure 9.5 Typical family configuration drawing with parameter callouts from The AlphaGraphics System (TAGS).

If a new variant is inescapable, it may, however, be possible to "make from" one of the existing designs.

Regardless of whether or not a new variant is required, it is an imperative that the system include a check point to review the design need. For this reason, the menu codes of those who create variety must not have the access to add or make permanent changes to the file. This point has been raised a number of times in other contexts, prior to this reference, and it is an absolute. The use of CAD could otherwise create a problem of cost consequences far in excess of its capability to reduce the cost of development and documentation.

The objective of design/process standardization must dominate all things that are done. That drafting can be drawn at the line rate of 2½ mph is a complete waste if the lines are for a part that should not enter the system in the first place. For this reason, the system that is recommended involves the variety controller to the same extent as in a manual system.

9.3 CAD/CAM AND PROPRIETARY DESIGNED PARTS AND ASSEMBLIES

Figure 9.6 is the diagram of a system that merges design variety control and CAD/CAM. Commencing with the layout, the designer conceives several possible solutions in the overview sense (i.e., mechanical, hydraulic, electrical, and/or combinations thereof). The designer then retrieves data using the family layout code for the several conceptual solutions from the data stored in the computer. This may be a listing of layout numbers for study, or actual graphical solutions—albeit, in a generalized sense. There are limitations that dictate the most reasonable way to keep the data. For example, a complete layout of a large electromechanical device drawn originally on a 12-ft roll size may not be practical to work with on a graphic CRT display that is 18 × 24 in. in size. Even if it is zoned, the overview will probably be lost to the designer. Whatever the solution to this problem, the data are visible in one form or another.

The designer then decides which of the several generalized solutions offers the best answer to the problem. If the answer is to use an existing layout "as is" without modifications, the problem is solved by the simple expedient of retrieving all part numbers for assemblies, piece parts, and purchased nonproprietary parts and assemblies, and enter them in the bill-of-materials file against the new model.

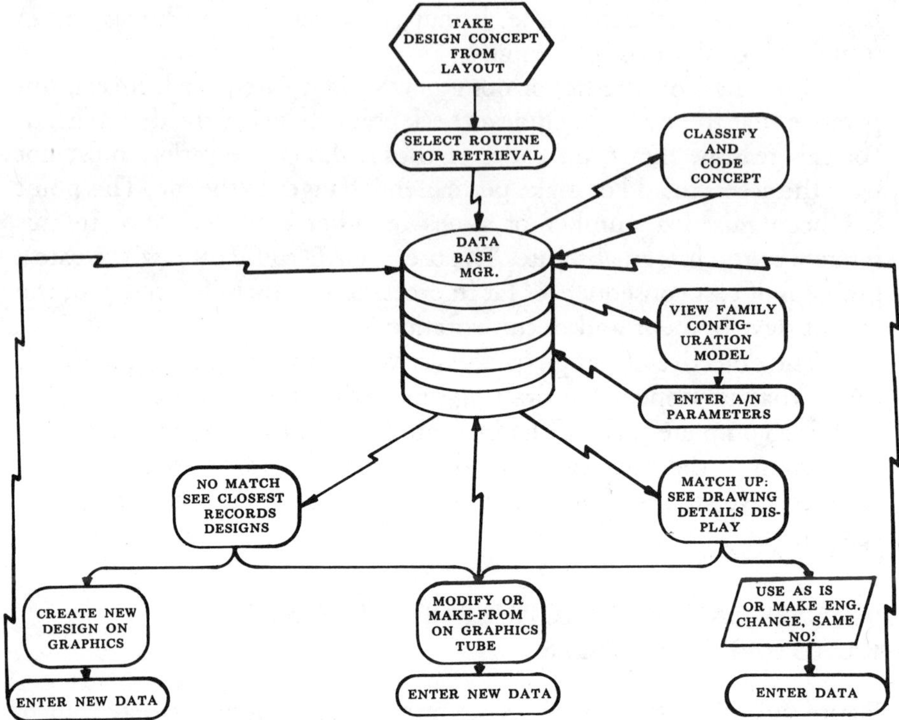

Figure 9.6 The AlphaGraphics System flow chart for enhanced computer-aided design.

If the exact solution is not found, but one very similar is located, the designer will redesign only the portion that must be altered. All other parts and assemblies will remain the same. If no close similar solution is found, the designer will create an original solution. This does not mean a license, however, to create new parts, assemblies, and nonproprietary parts and assemblies. The layout will utilize existing items to the maximum extent possible. The ingenuity to do this falls squarely on the designer and the draftsman, who will flesh out the layout design. As the need for visibility of designed parts and nonproprietary parts and assemblies arises, the data base will yield the family configuration and/or tabulate a listing as required. This is done on an item-by-item basis.

The part shown in Figure 9.7 is one of 61,000 parts distributed in 4950 families. The taxonomy is by part form, feature(s), and size.

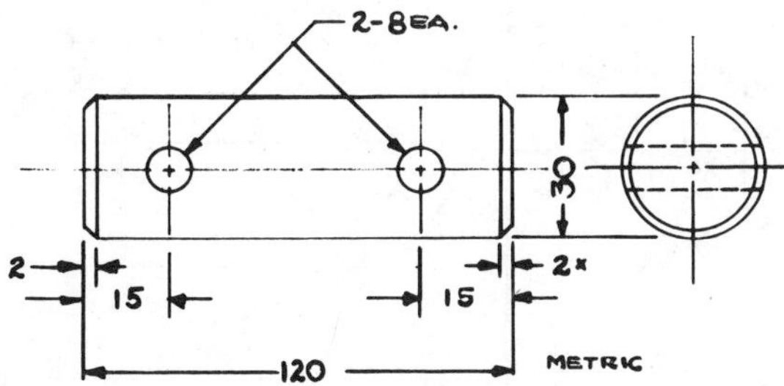

Figure 9.7 Part sketch for retrieval.

To find the appropriate family configuration model, the taxonomy (which has been digitized and stored) is presented graphically, as shown in Figure 9.8. It is in n-polar tree mode.

At this stage the appropriate form is chosen that most closely resembles the designer's concept. In this instance, group level code 301 represents single-diameter parts without a centerhole but with other holes. Group 301 is selected by light pen contact on the graphics tube or by cursor or tablet, as the system options dictate. This action continues the taxonomy (see Figure 9.9). These concepts can be reviewed singly and each blown up to full screen size if desired.

The design concept that has been the subject of the classification exercise most closely resembles the figure representing families 30170/30188, which embrace 370 individual parts. Because the code-number-growth capacity has been planned for the next 25 years, and 18,000 numbers have been reserved, it will be more than sufficient to accept this growth, even allowing for the influx due to metrication. Five hundred numbers in each family are reserved for U.S. Customary measured parts and 500 numbers are reserved for metric measured parts. If it were not for the need to number the parts, only one classification family would be required, in spite of the 370 parts the system accepts and processes—and the 18 families as one, in either case.

Another option for using the computer interactively is shown in Figure 9.10. It is n-polar expressed in NAND logic with built-in tutorials. The processor is part of The Alpha Graphics System.

With the appropriate configuration now on the graphics screen, the desired parameters are centered in the *designer's* order of priority: for example: (B) length, (C1 and C2) hole size, and material code. The designer has ignored certain parameters as not important to this design application.

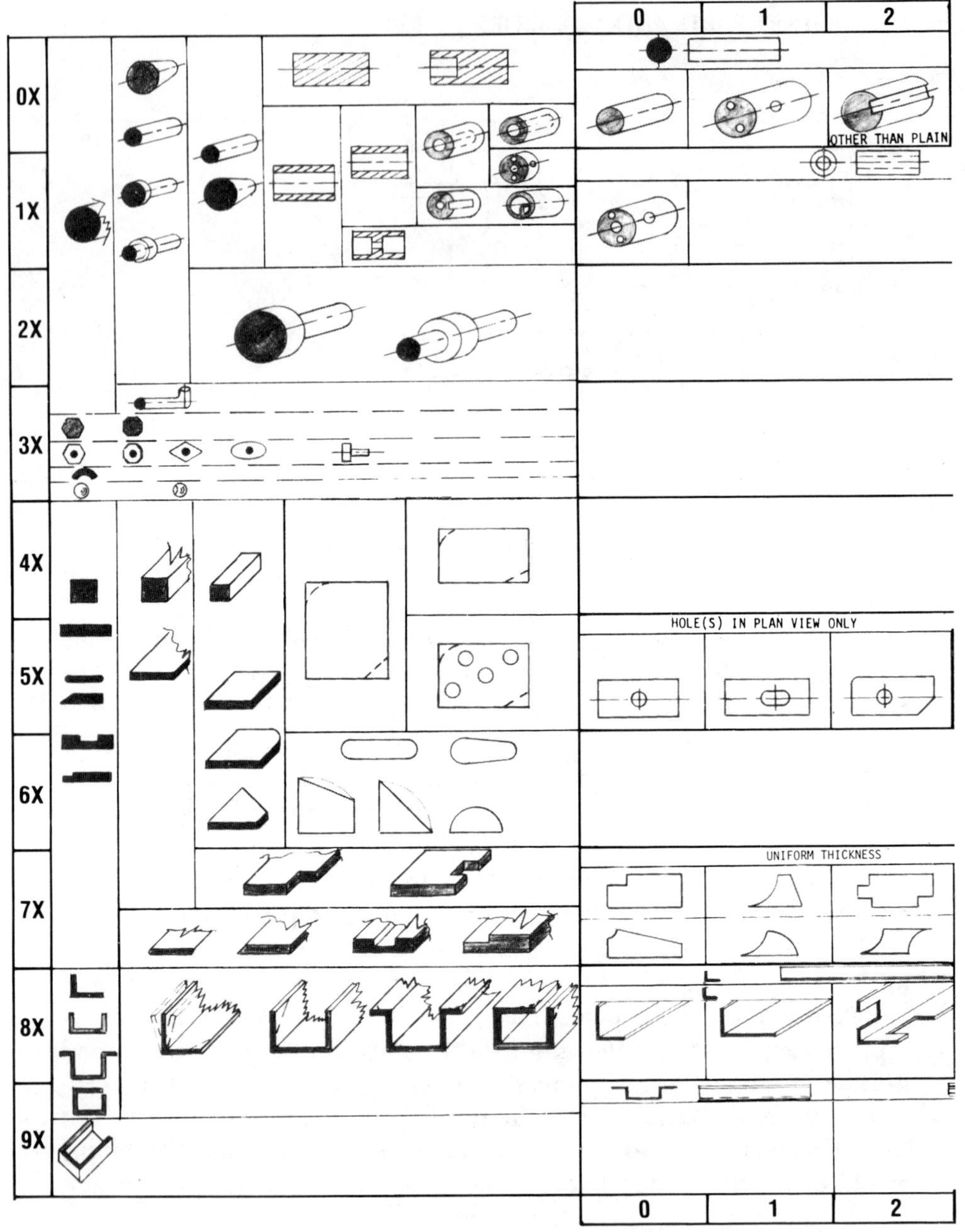

Figure 9.8 The AlphaGraphics System tailored group selection taxonomy on graphics set stored magnetically. (Copyright © Brisch, Birn & Partners.)

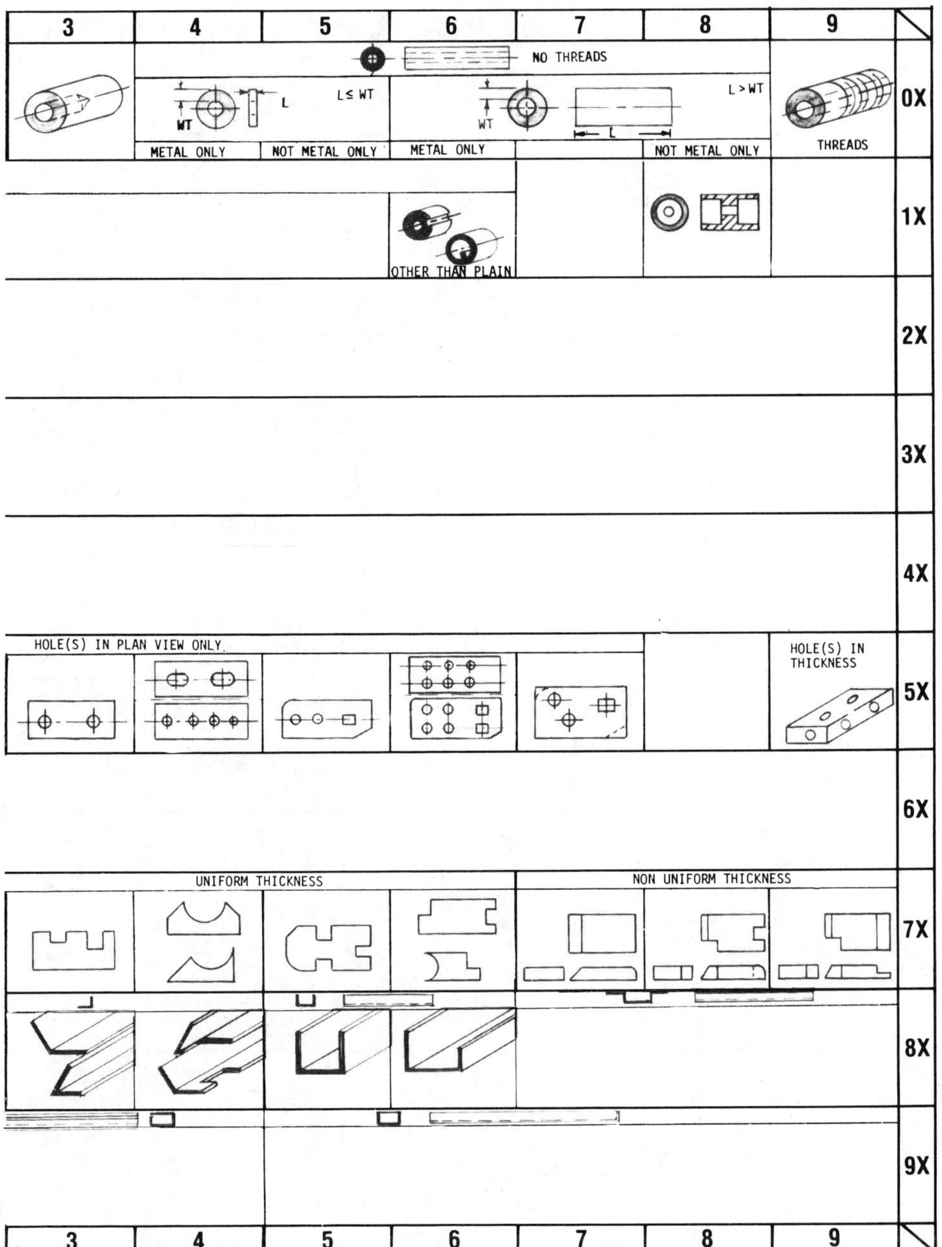
3
4
5
6
7
8
9
NO THREADS
L ≤ WT
L > WT
WT
L
METAL ONLY
NOT METAL ONLY
METAL ONLY
NOT METAL ONLY
THREADS
0X
OTHER THAN PLAIN
1X
2X
3X
4X
HOLE(S) IN PLAN VIEW ONLY
HOLE(S) IN THICKNESS
5X
6X
UNIFORM THICKNESS
NON UNIFORM THICKNESS
7X
8X
9X
3
4
5
6
7
8
9

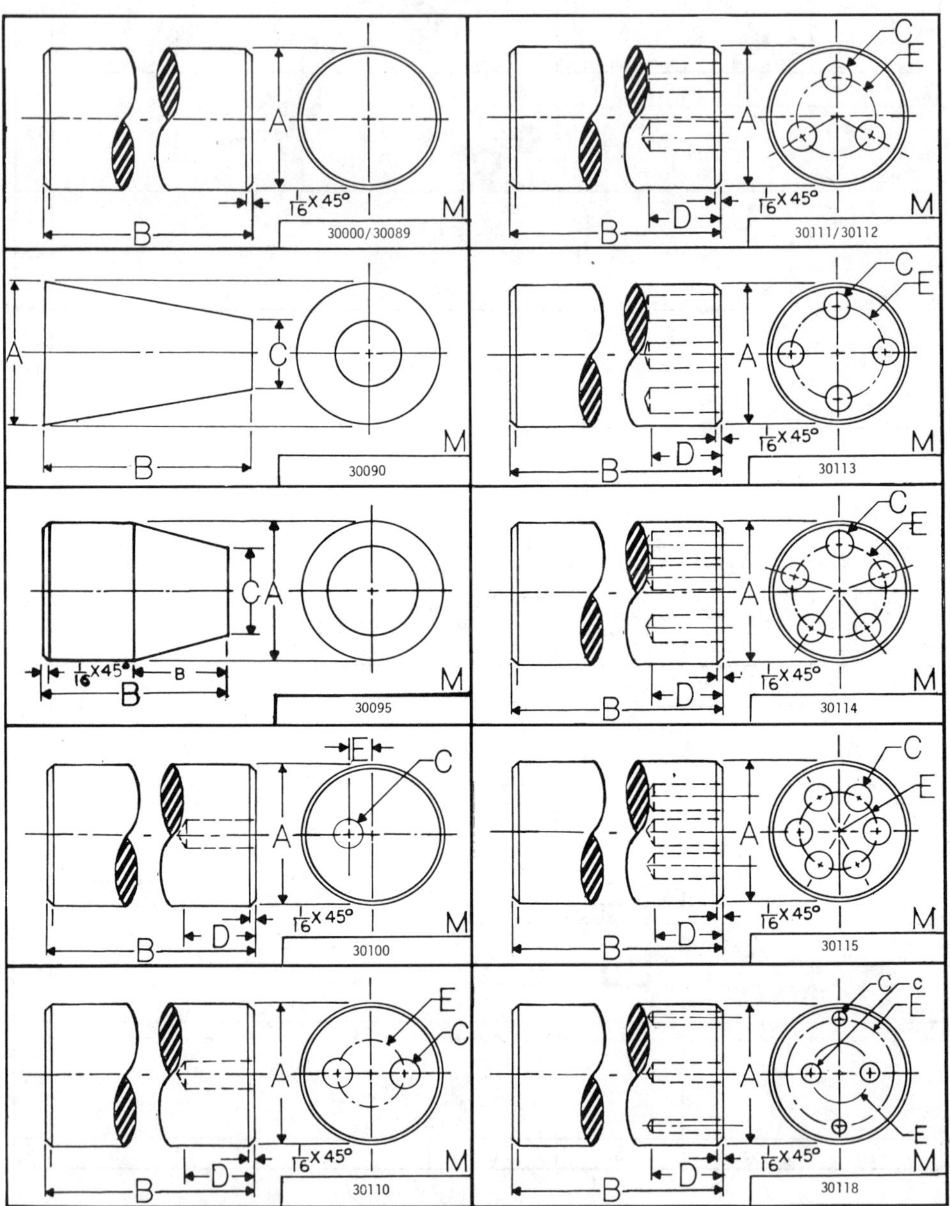

Figure 9.9 Family selection continuation of graphical taxonomy. (Copyright © Brisch, Birn & Partners.)

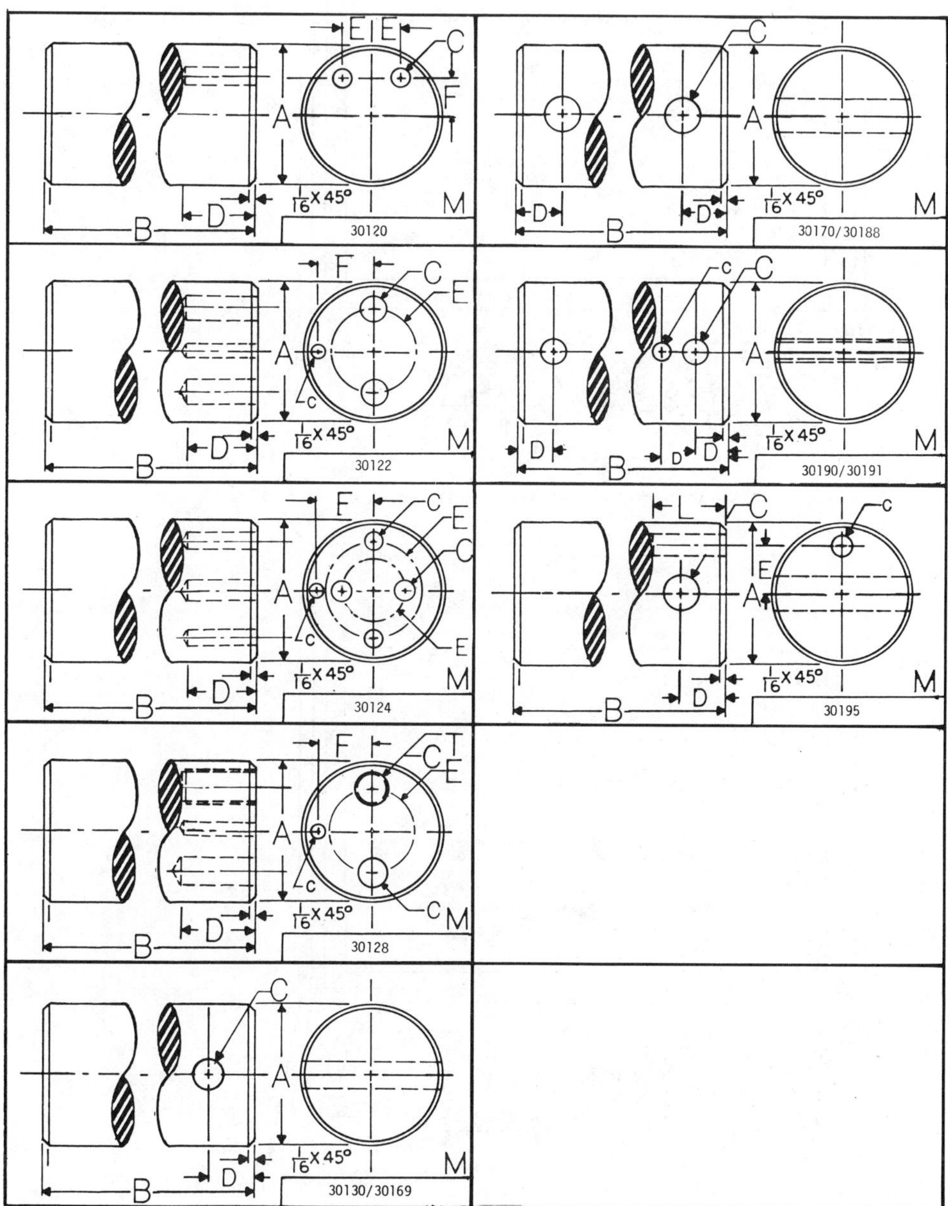
30120
30170/30188
30122
30190/30191
30124
30195
30128
30130/30169
1/16 X 45°

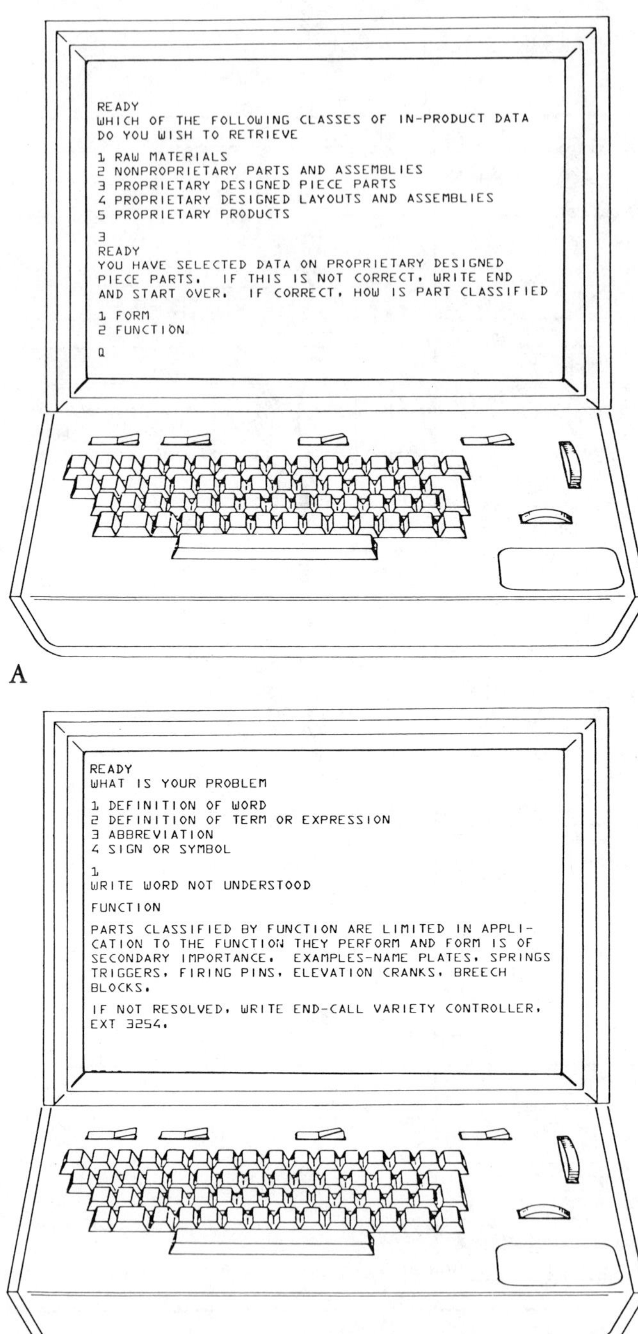

Figure 9.10 Interactive family selection option (TAGS).

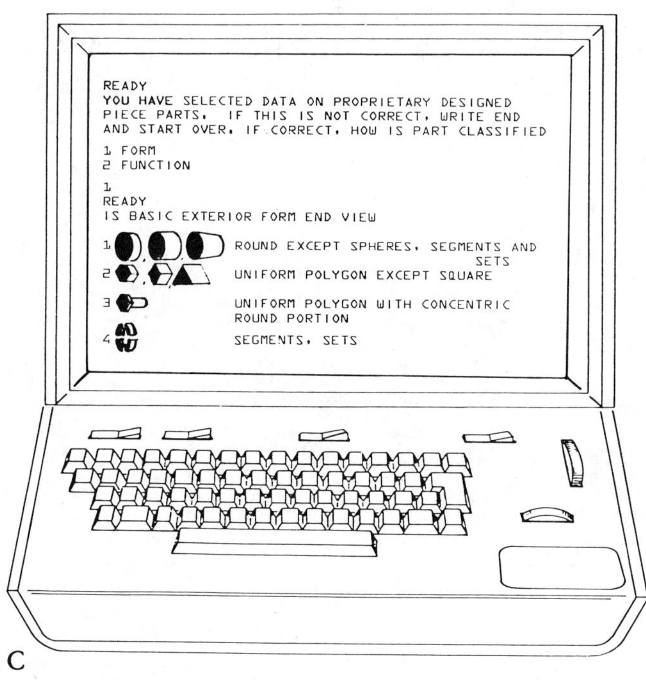

C

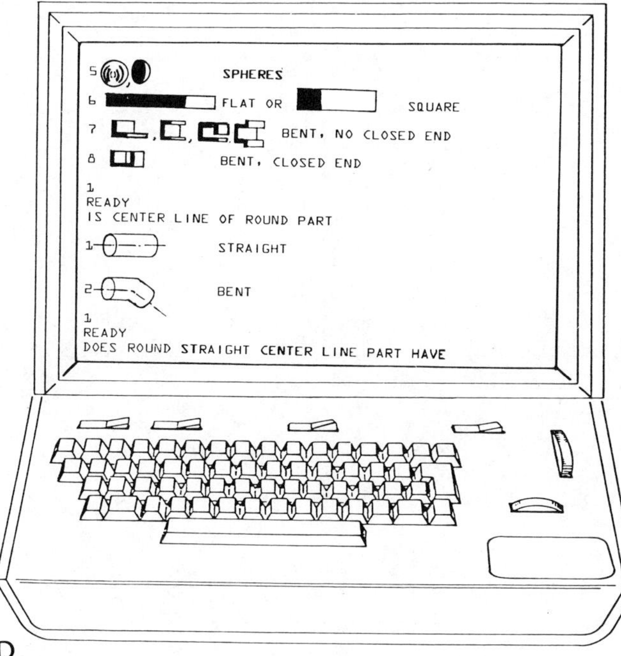

D

Figure 9.10 (continued)

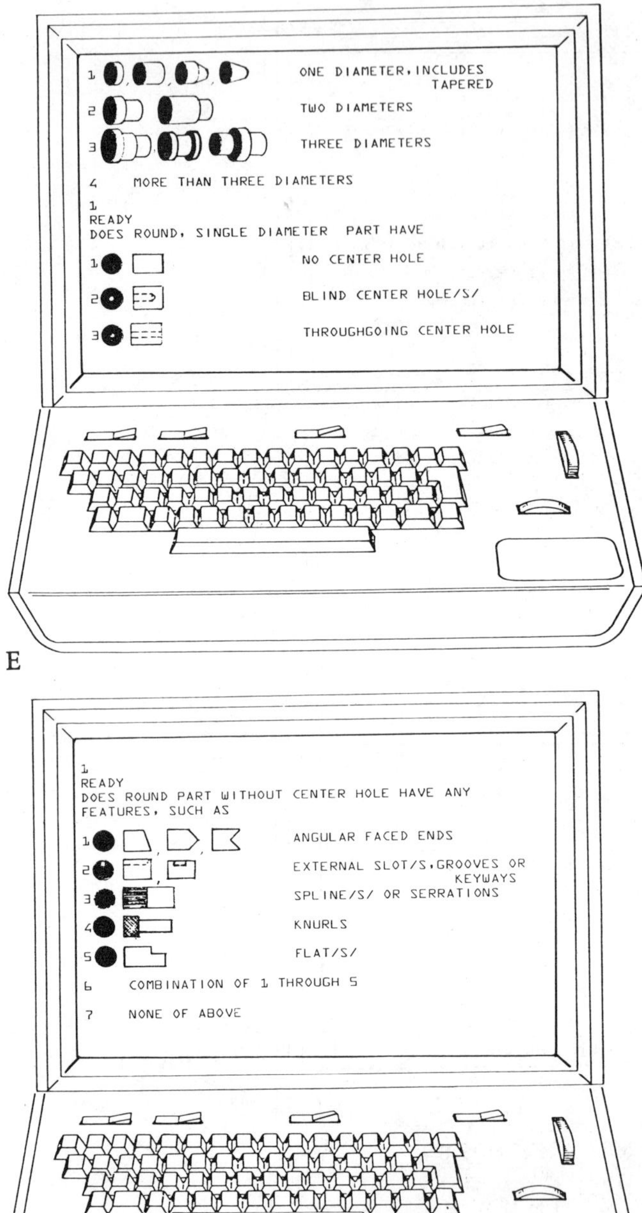

F

Figure 9.10 (continued)

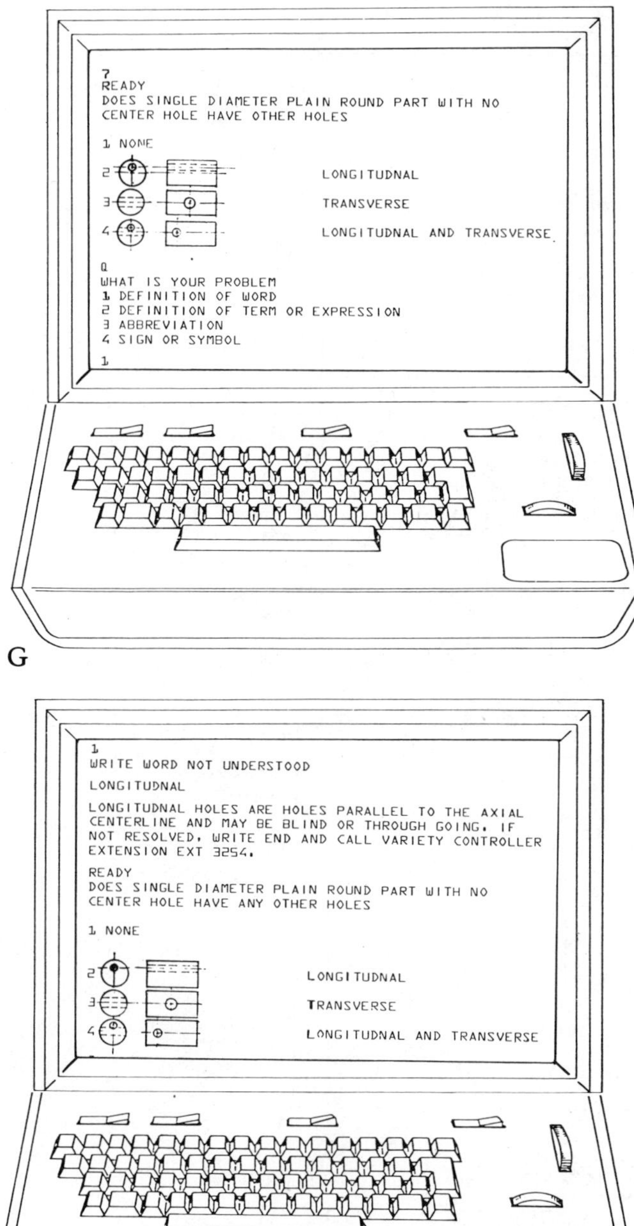

Figure 9.10 (continued)

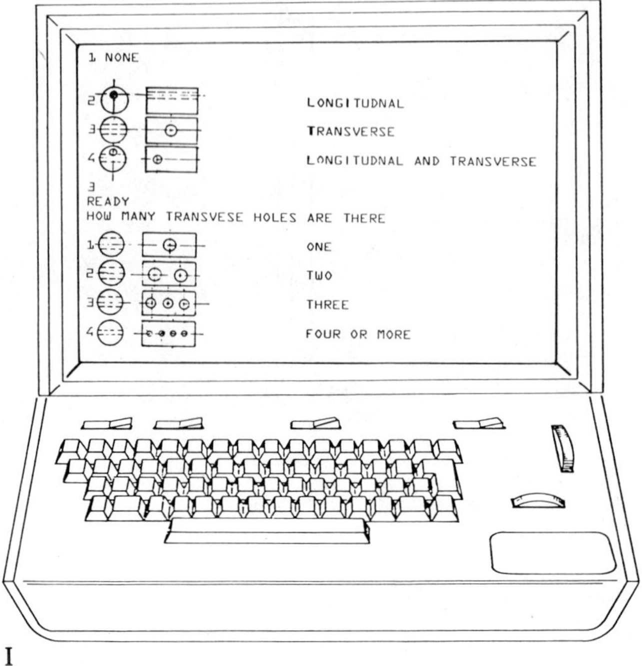

I

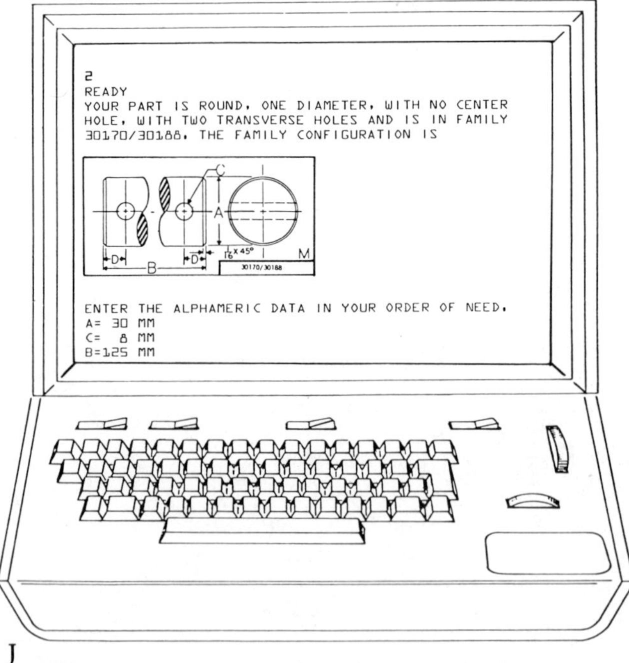

J

Figure 9.10 (continued)

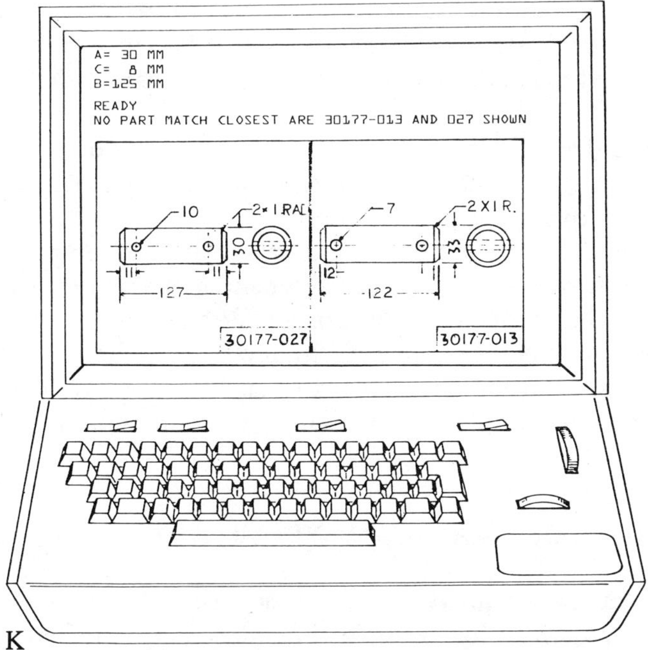

Figure 9.10 (continued)

The computer searches the parameters, which are stored in the data base or file, according to the system, in the order established by the designer. If there is a match, the graphic display of the actual part drawing will appear on the graphics tube. The designer then decides to accept or reject the existing part. If accepted, the existing code (part numbers) is added to the bill-of-materials file in the quantity required. If there are parametric differences (say, hole size, C1 and C2) that are not acceptable, the designer may elect to make an engineering change. That is, the designer will request a change if, after study of other applications, interchangeability is not affected. Suppose that the new design requires holes with tighter tolerances. The honing operations to achieve the new tolerances will not affect interchangeability, but the change request may be rejected because the honing operations costs are not offset by the lot size increase and other economies. If so, then obviously the decision is to make the new part from the old part. And, although a new part number will be created, all the advantages of lot size increase will be possible (if cost to carry inventory and/or scheduled needs cooperate) by making a combined requirements lot to the old number design.

The process routing for the new part will show "make from" and contain only the honing operations. The computer, however, will store the new design with all dimensions, materials, and other data added in the alphameric file.

A new part record and drawing are created. The system updates the old design to contain the changed or added data, complete with change level, notes, and the "make from" part number.

If, in the initial search, there had been no match up, the computer would exhibit the alphameric records of the lesser and greater parts. The process of analysis is then repeated as previously described.

When data of any kind are added to the file (or data base), it is flagged temporarily until reviewed within 24 hours by the variety controller. Only then will the temporary flag disappear and the number be made active.

9.4 RAW MATERIALS IN A CAD/CAM SYSTEM

The same equipment is used for raw materials selection. If a designer can use a part configuration, for example, but must have a different material because of the service environment, it is essential to select the materials from a specification preferred for future design. (The same procedure holds for selecting the raw materials for a new design.)

How this control if effected while using the computer to aid in the selection is represented by Figure 9.11. Here the designer needs to see all resulfurized steel round bar stock. The classified and coded specifications are stored for engineering selection. The computer terminal is used to access the family data by entering the code found in a computer-stored index from a hard-copy tutorial, or by interaction (as explained in Chapter 8 and shown in Figure 9.10 for piece parts).

Because it is at the designer or draftsman terminal, accessing the computer with the appropriate menu code will bring up only preferred specifications. These specifications show material chemistry and the key dimension of the materials preferred for future design.

The file of information, however, contains all specifications, not just preferred specifications. It also contains the purchasing specifications, which include all data necessary to describe the actual items to be bought. Length of bar, for example, is included in the purchasing specification, although it is not shown on the engineering display.

The designer enters 12210, the code number for the family of round, resulfurized, low-carbon bar. Relevant heading data, including standards references applicable, type, form, and the like, appear. Also on the display are column headings that show the register for entering data for the particular set of parameters and the recommended uses. The column in which status is shown in this illustration, P for preferred, tells the designer which specifications shown are available for use in future designs (the program is written to display only preferred items at the design terminal).

Figure 9.11 CAD/CAM display for selecting raw materials specification. (System copyright © 1976 Brisch, Birn & Partners.)

There are two specifications that can be used. That raw material specification number is used and is added to the material block in the title. This ends the search.

If there is no suitable specification (chemistry and size), the designer or draftsman enters the needs on the tutorial cathode ray tube (CRT) terminal, according to the instruction given. The printer will print out the request and repeat at the variety control terminal. The variety controller then communicates with the designer by computer, voice box, or telephone. If the designer can be convinced to use another chemistry of steel that is preferred and/or another size larger that is preferred, that ends the dialogue. If not, and there is good and sufficient reason, the variety controller will, from a full display of all specifications, provide a nonpreferred code number that satisfies the need if one exists. If there is no specification, preferred or nonpreferred, the variety controller will assign a part number to the now new specification and enter it into the file. However—and this is a very important qualification—the new item will not be given a preferred status. It will be made nonpreferred until such time as there are enough demands for it to be reviewed for status change. This means that even though there was a sufficient argument given for one application, it will not become visible to other designers until such time as it is upgraded to either *preferred* or *standard* status. In this way, an effective control can be maintained over the variety of raw materials while not handicapping design efficacy.

The actual raw material item to be purchased (see Figure 9.12) will then be selected by a manufacturing engineer and/or production or inventory controller. One of the parties assigned the responsibility will prescribe the most appropriate length for the part, taking into consideration the order size for the part in question. Obviously, a 10-ft shaft should not be made from 8-ft random-length stock. Neither should a part 25 mm long be made from a 5-m random bar, unless, of course, the order calls for 100 or more.

9.5 NONPROPRIETARY PARTS AND ASSEMBLIES SELECTION WITH A CAD/CAM SYSTEM

A mode of operation similar to that used to select raw materials aids the designer in selecting nonproprietary parts and assemblies. There is one fundamental difference. These items require support documentation. The rate of change in supplier products requires that documentation be

Figure 9.12 CAD/CAM display of purchased raw materials to specification.

updated to reflect changes in the maker's part number. Industry and national standards are also updated from time to time. These reference changes must be reflected in the supporting documentation, as well. For this reason, it is useful to store and display all such data that might be affected by these changes. The variety control station should be used interactively with editing capability to update the data.

How the computer is used in a CAD/CAM system to aid the designer and/or any other user to select a preferred nonproprietary part or assembly is reflected in a flow chart of a system. In Figure 9.13 the selection process is shown as well as the file update procedure. Metrication will accelerate this aspect of the system, and will require frequent consulting of engineering materials and standards people.

The access to the computer for the family or families of items is very much the same as for raw materials selection. If, for example, the designer wishes to select a ball bearing, single row, Conrad type of a cer-

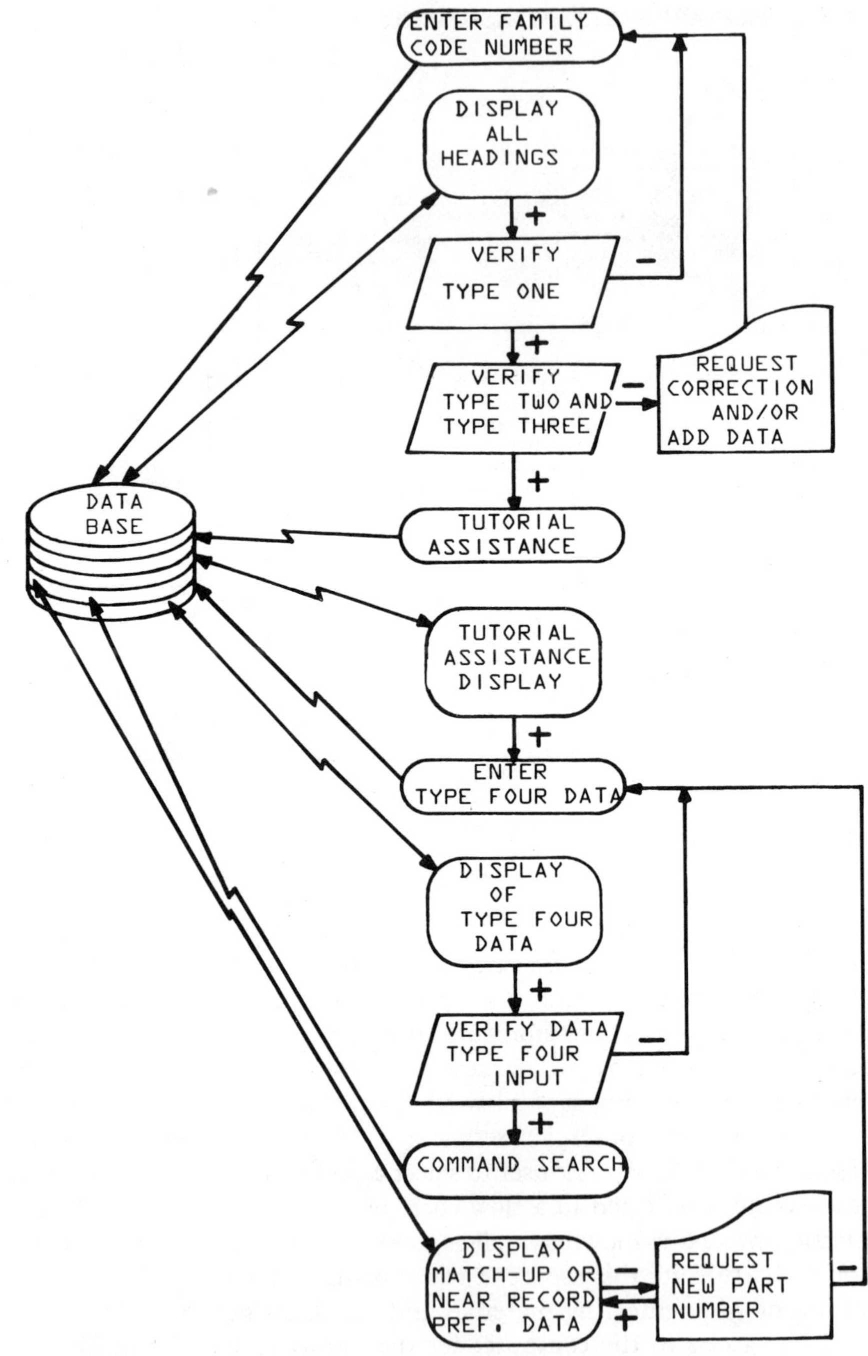

Figure 9.13 Flow chart of retrieval on the CAD/CAM system of materials and/or nonproprietary parts and assemblies.

	FAMILY NUMBER
BALLS	
Bearing	2290
BANANA PLUGS	2774
BANDING CLIPS (SEALS)	2802
BANDING TOOLS	
Pliers	287
Tensioners	2897
BASES	
Motors	2231
BEARINGS BALLS	2290
BEARING LOCKNUTS	2295
BEARING LOCKWASHERS	2296
BEARING NEEDLES	2291
BEARING ROLLERS	2291
BEARING SHIMS	2293
BEARING SLEEVES	2298
BEARINGS	
Ball	
Single Row Including Duplex	
Without Extended Inner Race	2270
Double Row	2273
Impregnated	
Unmounted	
Plain Sleeve	

Figure 9.14 Hard-copy alphameric index. (Copyright © Brisch, Birn & Partners.)

tain bore size and accuracy (ABEC accuracy code number), the file may be accessed for the family code from any of the several tutorials:

From a hard-copy index of these items (see Figure 9.14)
From the same information stored as a tutorial
From the same data programmed for access by a key word noun and descriptors
By interaction (see Figure 9.10)

Once the family information desired—headings and column headings—are visible on the graphics CRT, the designer may enter the specific alphanumeric data for the specific bearing desired.

Figure 9.15 shows the family description and data requirement for the type of bearings needed. There is a problem here in that the data needed require more columns than the usual CRT (80 columns, 26 lines) is capable to display. As referred to in Chapter 8, there are alternative ways to cope with the problem.

Regardless of the solution in the hardware, for a proper selection, the software must present all the data, one way or the other. Presuming that this has been resolved, the parameter information, once entered, will either match an existing item preferred for future design, or it will not. If there is no match, the printer will print out the parameters of the closest (smaller and larger) preferred item similar to the designer's requested item. Thus, the designer can see which of the parameters do not match and decide either to accept one of the alternatives or not to.

If the designer decides that a new item is needed, the system causes the printer at the variety control station to request a new part number. As in the raw materials example, a dialogue will then ensue between the variety controller and the requesting designer (or materials manager, or whoever the requester may be). The dialogue may be by computer, telephone, or voice box, according to the conditions involved and hardware employed.

If a persuasive argument and a justification can be given to substantiate the need for a new item, the variety controller will either:

Assign an existing nonpreferred item number from the file, or
Assign a new part number, and enter the new data parameters into the file.

The new item, however, will print on the printer at the designer's terminal but will not be visible to other users, as it will be a nonpreferred item. If other demands for the same item occur, a review will be made to determine if a change in status is required. This prevents proliferation of unnecessary variety and maintains the posture of control essential to a proper product engineering standards program.

9.6 MINI VERSUS MACRO VERSUS MAXI COMPUTER-BASED CAD/CAM SYSTEMS

The stand-alone dedicated mini CAD/CAM systems created the market and currently dominate it. Their successes to date have market analysts predicting exponential growth for CAD/CAM technology. The sweet

BEARINGS
BALL, SINGLE ROW, INCLUDING DUPLEX, WITHOUT EXTENDED INNER RACE
RADIAL RIGID
UNLESS OTHERWISE SPECIFIED

FAMILY 2270
PAGE 2
DATE 04/06/79

CATALOG LISTING

SECTOR RANGE 000 THRU 999 ALL

SEC	NOMINAL BORE DIA MM	NOMINAL OUTSIDE DIA MM	OVERALL WIDTH MM	NOMINAL BORE DIA INCH	NOMINAL OUTSIDE DIA INCH	OVERALL WIDTH INCH	NUMBER OF SHIELDS	NUMBER OF SEALS	OTHER DATA	PREF	SEC	LOCATION HAZ	LOCATION ALN	LOCATION STM	LOCATION IPR	PART NUMBER	F/M
121	60	110	22	2.362	4.331	0.866			SKF 6212	N	121	X				X08-1235	
183	60	110	22	2.362	4.331	0.866			SKF 6213		183					X08-1033	A
143	60	110	36.5	2.362	4.331	1.438			SKF 5212	P	143	X				01A10741-01	
103	60	130	31	2.362	5.118	1.221			MRC 7312 D ANGL CONT	P	103	X				01A16446	
101	65	120	46	2.559	4.724	1.811		1	MRC 213 SZDB	P	101	X				01B11003	
127	65	140	33	2.559	5.512	1.299	1		SKF 6313 Z	N	127	X				X08-1040	
155	65	140	33	2.559	5.512	1.299			CLASS-ABEC-1	P	155	X				15D8291	A
						MATCHED SET CONSISTING OF MRC-9313-UD & MRC-7313-D											A
033	65	140	66	2.559	5.512	2.598		1	DUPLEX MRC 7313 DT	N	033	X				01A16538	
169	65	140	66	2.559	5.512	2.598			MRC-7313-DB-ABEC-1	N	169	X				1517	A
107	65	160	37	2.559	6.299	1.457			SKF 7413 B ANGL CONT	N	107	X				X08-1099	
									FAFNIR 7413 PW								
117	70	125	24	2.756	4.921	0.945			SKF 6214	N	117	X				X08-1244	
125	70	125	24	2.756	4.921	0.945	1		SKF 6214 Z	N	125	X				X08-1186	
007	70	150	35	2.756	5.906	1.378	1		SKF 6314 Z	N	007	X	X			X08-1041	
123	70	150	35	2.756	5.906	1.378			SKF 6314	N	123	X				X08-1187	
005	70	180	84	2.756	7.087	3.307			SKF 7414 BG ANGL CONT	N	005	X				X08-1191	
067	75	115	20	2.953	4.528	0.787			SKF 6015/C3	P	067	X				01A11075	
069	75	130	25	2.953	5.118	0.984			SKF 6215/C3	P	069	X				01A11074	
157	75	130	25	2.953	5.118	0.984			MRC-215-S-ABEC5	P	157	X				73150	A
053	80	140	26	3.150	5.512	1.024			SKF 6216	N	053	X				X08-1091	
111	80	170	39	3.150	6.693	1.535			SKF 6316	N	111	X				X08-1094	
109	85	180	41	3.347	7.087	1.614			SKF 6317	N	109	X				X08-1095	
087	90	160	30	3.543	6.299	1.181	2		MRC 218 SFF	P	087	X				01A16352	
099	90	160	30	3.543	6.299	1.181			SKF 6218	N	099	X				01A12708-01	C
									MRC 218 S								
193	90	160	30	3.543	6.299	1.181			MRC 218-S (ABEC-5)	N	193	F				73137	A
129	90	160	60	3.543	6.299	2.362			SKF 7218 BG	P	129	X				01A12650-01	
057	90	190	43	3.543	7.480	1.693			SKF 6318	N	057	X				X08-1067	
									FAFNIR 318 K								
137	90	190	86	3.543	7.480	3.386			MRC 7318 DU	P	137	X				01A14787-01	
051	95	200	45	3.740	7.874	1.772			SKF 6319	N	051	X				X08-1069	
045	100	215	47	3.937	8.465	1.850			SKF 6320	N	045	X				X08-1234	
187	100	215	47	3.937	8.464	1.850			SKF 7320		187					X08-1050	A
089	105	190	36	4.134	7.480	1.417	2		MRC 221 SFF	N	089	X				01A16353	
181	105	190	36	4.134	7.480	1.417	1		SKF 6221 Z		181					X08-1035	A
091	130	200	33	5.118	7.874	1.299			SKF 7026 C/C78, ANGL CONT	N	091	X				X08-1026	
149	130	230	40	5.118	9.055	1.575			FAFNIR 7226WO-MBR-SU-FS	P	149	X				01A14174-01	
147	140	250	42	5.512	9.843	1.654			SKF 6328	P	147	X				01A14173-01	
173	150	225	35	5.906	8.858	1.378			SKF 7030CTC/C78	P	173	X				X08-1188	A
									ANGL CONT								A
003	150	270	45	5.906	10.630	1.772			SKF 7230 ANGL CONT	N	003	X				X08-1189	
171	165	216	25	6.500	8.500	1.000			KG65AH5DB-ABEC-4	P	171	X				01A11261-01	A
075	220	340	56	8.661	13.386	2.205			MRC 144 KR	N	075	X				X08-1209	
077	220	340	112	8.661	13.386	4.409			MRC 144 KRD	N	077	X				X08-1209	
035	320	412	38	12.598	16.220	1.496	1		FAFNIR 264K2BR	N	035	X				X08-1124	
									TORRINGTON D2439-B								
175	762	889	64	30.000	35.000	2.500			TORRINGTON D25738 (SPL)	P	175	X				X08-1220	A
									ANGL CONT								A

FAMILY 2270

Figure 9.15 Retrieval of ball bearings in a system. (Copyright © 1977 Brisch, Birn & Partners.)

scent of a $2,000,000,000 plus potential by 1985 for these kinds of systems is attracting the big bees to the marketplace.

IBM has already adapted its 4300 Series macro computers to CAD/CAM application. CADAM, originally conceived and used on large-scale, large-data-base computers, resulted from a collaboration with Lockheed Aircraft some years ago. Control Data Corporation, in 1978, made a major commitment in funds and personnel to develop a large-scale-integrated system to support as many using functionals' requirements as any one firm might find profitable. This system is now operable and available. It is large, powerful, versatile, and modularized. A key element of the system is data retrieval through the use of a classified and coded data base, as illustrated by the flow chart, Figure 9.4. Because of its modular concept, as needs justify, modules may be added. As more users within the firm need to access the data base, more terminals may be added without interference, tube flicker, signal diminution, and the like. Time sharing is available, as well, for those whose computers may not be able to accept the specific requirements.

The most important contribution of this new entry is that it integrates the cycle of business and scientific activities, including:

Product research and development
Product design
Product engineering
Product standardization
Documentation and configuration control
Manufacturing engineering
Industrial engineering, and
Tooling

and accommodates growth. The engineering/manufacturing interface is dramatically improved, as each has free access to the same data base.

Whether to choose a mini, a macro, or a maxi system is not easily resolved. Perhaps the day is coming when the stand-alone mini/macro computer systems now in place will become super smart terminals linked to the larger-scale control processing units. Several of the mini system makers see this opportunity and have begun to work on the premise that compilers for language compatibility are possible.

What should your firm do? Only you can really say. Large and small are relative and in the eyes of the beholder. It is not the purpose of this book to make your decision for you. There are, however, some key factors that you might consider when you make your study:

How many inquiries are there likely to be to the system per hour?
How dispersed geographically are the inquirers?
How much storage will be required to accommodate all the active items data that will be in the file?
Will tool design and gage design data be included?
What is the average life of parts, assemblies, and materials data active in the system?
What is the maximum interference acceptable for terminal availability?
What is the maximum response delay time that you will accept?

Answers to these questions will provide some clue as to the size of the data base and the number of terminals that will be needed to get started. To this will have to be added your growth rate and pocketbook.

Your plan must be an orderly means to accept growth and increase in CAD/CAM activity without extensive revision and cost in the future. Regardless of what you decide to do, there is a CAD/CAM system in your future—if not today; and when you have made your decision and chosen your system, make certain that:

The system is adaptable to your requirements and needs; and not the reverse.
The system has a foolproof means to control variety by retrieval of prior data.
The reporting system of actual versus planned is looped back to those who funded the project.
The system recognizes those who use the system effectively and produces results.

9.7 ALL THAT GLITTERS . . .

There are a number of problems that must be addressed. Not everyone is suited to sit in front of a computer terminal day in and day out. A study has just been made by a midwestern medical school on a government grant to investigate problems and formal complaints made regarding terminal use.

The preliminary results ruled out radiation leaks. But the study did reveal a number of psychological and physiological problems that prolonged and unbroken terminal use evokes. Testimony and case history documentation were studied. What the final report will say will not be known for some time, but the inference is that not all temperaments are comfortable working at terminals for prolonged periods. Part of the problem is psychological; specifically named was the effect of working in a darkened room.

Even before these preliminary findings were aired, one firm had experienced problems and attempted to find out why some designers were successful and others were not. This firm, for example, found that only 11% of their people could design in three dimensions. Attempts to find out why this situation existed turned up the following factors:

Optical deficiencies, especially in senior designers.
Temperament differences in spatial relationships' conception.
Job conditioning in orthographic projection.
Fear of computers.
Resistance to change, especially among the more experienced and senior designers.

This firm trained all who could make use of the system. They selected the most successful of the trainees as design group leaders. They reorganized their 100 plus designers and draftsmen into groups of approximately 10 each, led by a CAD terminal expert.

In many instances, senior designers were subordinated to people many years their junior. Obviously, this spawned other problems and some good people could not adapt. The point here is that this firm felt that their future and CAD/CAM were so linked that they were willing to lose a few experienced and talented people.

On balance, the net result offset the discontinuities and disgruntlements. When asked if, with all the problems experienced, given it to do again they would have installed CAD/CAM, the answer was an unqualified "yes."

9.8 GENERATIVE PROCESS PLANNING AND CODING

There has been significant progress in the field of generative process planning (i.e., file search for closely similar parts and processes). Most are restricted to form classified parts.

Figure 9.16 is a system which includes generative process planning using a CAD/CAM system. It is significant in that the total product design and engineering problem and the total manufacturing engineering, tooling, and industrial engineering problem are addressed to find a common solution.

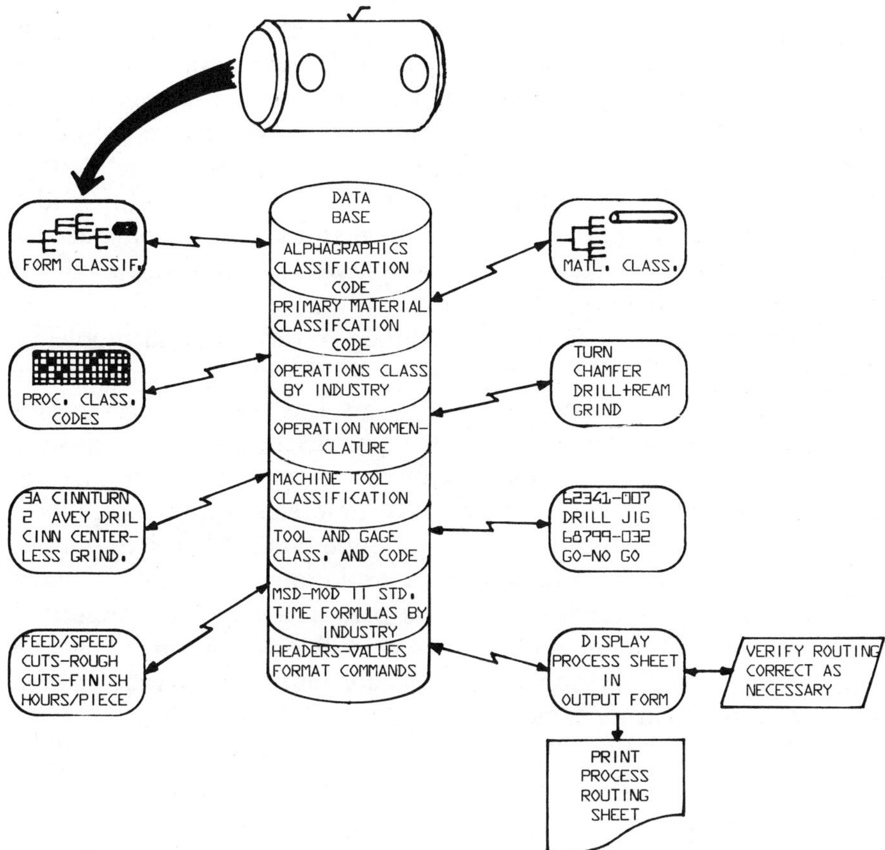

Figure 9.16 CAD/CAM system flow chart for preparing a process by interactive retrieval of standard sets of data in a data base (The AlphaGraphics System®, Brisch, Birn & Partners.)

Anyone who has tried to solve the technical and human interface problems between product design and manufacturing is well aware of the difficulty. Some firms have gone so far as to use a project approach to new design, incorporating teams of designers, draftsmen, manufacturing engineers, and industrial engineers to overcome the communication and organizational roadblocks.

As shown in Figure 1.6, the first computer-aided time standards were developed by IBM Corporation and Serge A. Birn Company. The system, called AUTORATE, was available to the public in the early 1960s. The only problem was timing—the computer was too new and few industrial engineers could be persuaded to use a computer to establish time standards, let along processes. However, to provide a common data base that does not distinguish between the departmentalization inherent in the management structure is the future and is possible now. This is perfectly correct, because how a firm is organized is a nonpermanent characteristic. A total systems approach should, therefore, logically ignore organizational bias.

This system proposes a single solution to the manifold problems. The end result of the design aspect is a computer-generated image representing the part. The manufacturing solution is a process to produce the part as designed; the tools required; the data to control the equipment; time, cost; and feedback to evaluate performance. Regardless of whether it is found necesary to interact with the computer to classify and code or not is of secondary importance. What is important is the unified design and manufacturing solution.

In summary, there is no ready, or pat, answer to the question: "I see how it works, but our work is different; will it work for us?" The only way to answer that question is to investigate your own situation or have an independent authority do it.

10

Classification and Coding and Manufacturing Planning

10.0 SCOPE

To look at the total problems of maintaining control of production and planning effectively for its accomplishment, it is necessary to look at other populations of data. As shown in Figure 10.1, these include such items as:

Maintenance, repair, and operating supplies (MRO)
Machinery, equipment, and supporting makers' proprietary spare parts
Perishable tools
Part-specific tools, restricted in application

These items, although not part of the end product, are essential to the manufacture of the product. If a critical machine is down on a progressive line because a $30 bearing seized, the entire line is down. For this and other reasons, it is most useful to have these data visible on a readily available spare bearing.

The expenditure for these items is a significant factor of total cost. This is so even though much of the cost is indirectly applied as general factory overhead and largely invisible. Exceptions are the part specific tooling limited in application to a part, assembly, and/or a family of parts. These are charged to the product.

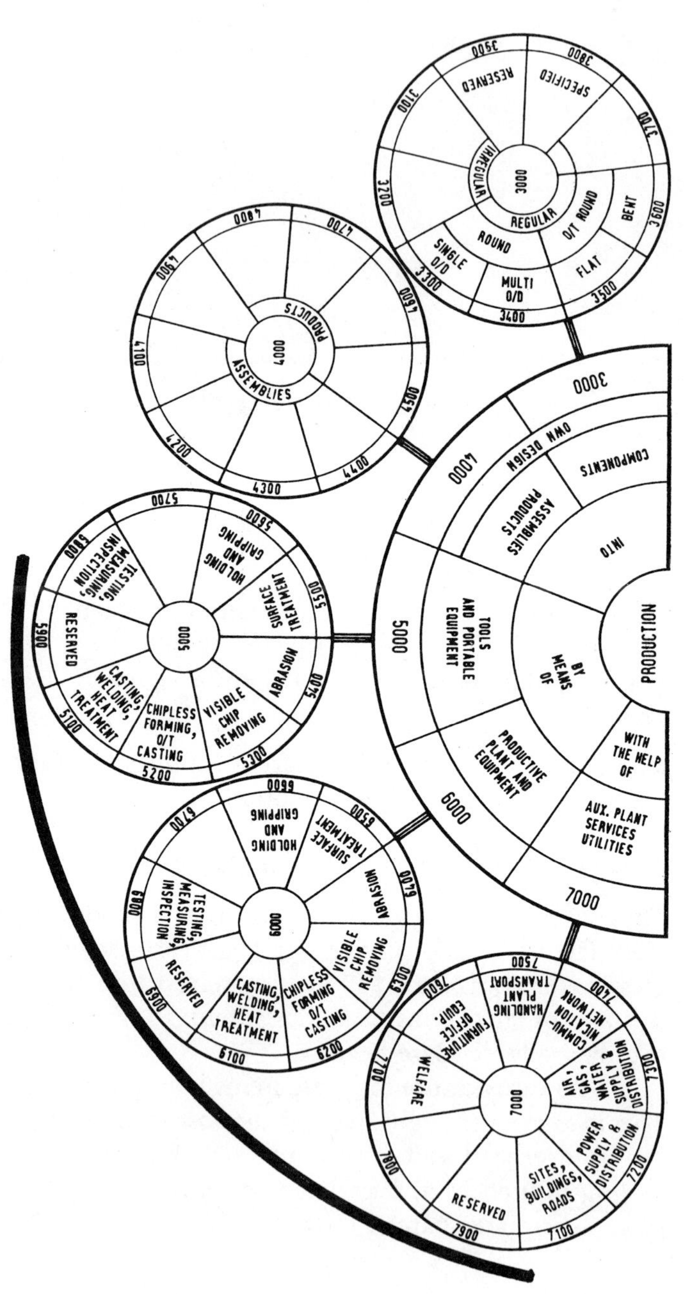
PRODUCTION
INTO
BY MEANS OF
WITH THE HELP OF
COMPONENTS
OWN DESIGN
ASSEMBLIES PRODUCTS
TOOLS AND PORTABLE EQUIPMENT
PRODUCTIVE PLANT AND EQUIPMENT
AUX. PLANT SERVICES UTILITIES
3000
4000
5000
6000
7000
3000
REGULAR
IRREGULAR
ROUND
O/T ROUND
BENT
FLAT
MULTI O/D
SINGLE O/D
SPECIFIED
RESERVED
3100
3200
3300
3400
3500
3600
3700
3800
3900
4000
ASSEMBLIES
PRODUCTS
4100
4200
4300
4400
4500
4600
4700
4800
4900
5000
CASTING, WELDING, HEAT TREATMENT
CHIPLESS FORMING, O/T CASTING
VISIBLE CHIP REMOVING
ABRASION
SURFACE TREATMENT
HOLDING AND GRIPPING
TESTING, MEASURING, INSPECTION
RESERVED
5100
5200
5300
5400
5500
5600
5700
5800
5900
6000
CASTING, WELDING, HEAT TREATMENT
CHIPLESS FORMING O/T CASTING
VISIBLE CHIP REMOVING
ABRASION
SURFACE TREATMENT
HOLDING AND GRIPPING
TESTING, MEASURING, INSPECTION
RESERVED
6100
6200
6300
6400
6500
6600
6700
6800
6900
7000
SITES, BUILDINGS, ROADS
POWER SUPPLY & DISTRIBUTION
AIR, GAS, WATER & SUPPLY & DISTRIBUTION
COMMU-NICATION NETWORK
HANDLING PLANT TRANSPORT
FURNITURE OFFICE EQUIP.
WELFARE
RESERVED
7100
7200
7300
7400
7500
7600
7700
7800
7900

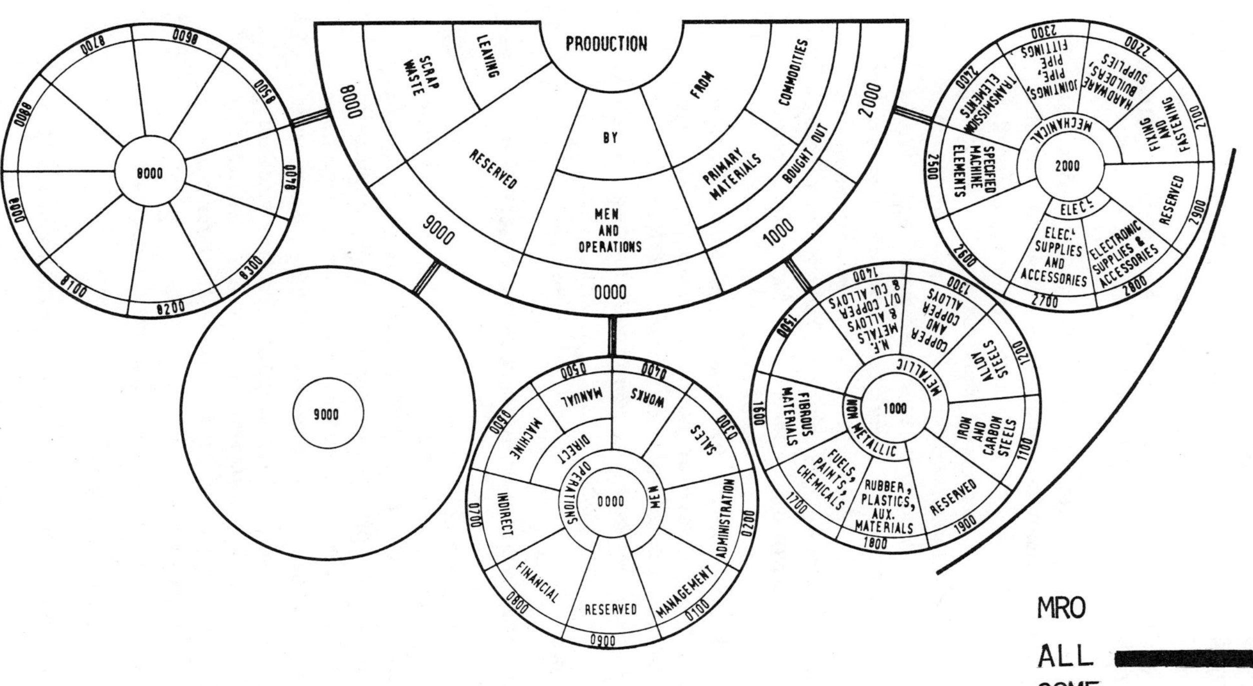

Figure 10.1 Other data influences on manufacturing planning.

10.1 DEFINITIONS

The definitions of the several populations of the four categories preceding are as follows:

> *MRO.* These are items that you buy to aid and support production which are used in the manufacturing process but do not become part of the product. Included are also the materials of maintenance, construction, and repair. *Examples: Janitorial and housekeeping chemicals, cleaners, fluid absorbers for spills, and accessories such as mops, brooms, waste containers, and dust pans; wipes, toilet tissue, toweling, and rag waste; preventive maintenance materials and accessories, such as hydraulic fluid, lubricants (oils and greases), lubrication guns, and oil cans; construction items, such as floor patching compounds, cement, bus bar, wire, insulation—thermal and electrical; structural steels, rod, bar, sheet, and plate, pipe, tubing, conduit, re-bars and wire mesh; unrestricted application spare parts, such as fasteners, bearings, bushings, capacitors, resistors, terminals, electric motors, generators, inductors, starters, reactors, junction boxes, and conduit fittings; chokes, valves, cocks, pipe and tube fittings, metering devices, and hydraulic motors.*

The list is virtually infinite and usually exceeds the variety of items purchased for end-product inclusion.

> *Machinery, equipment, and supporting makers' proprietary spare parts.* These are the machinery and equipment that comprise the production (and support) processes to manufacture the product, including accessories and handling equipment. *Examples: machine tools, machine tool accessories, material handling, hoists, lifts, trucks, conveyors, container racks; strapping machinery, stapling machinery, and packaging; coil winders, dereelers, and washing machinery; and the proprietary makers' spare parts for replacing damaged and/or worn parts to maintain as closely as possible the process capability for which the equipment was purchased.*

This list, too, can be quite extensive.

> *Perishable tools.* These are items that are bought (or made) that remove material and/or contribute to its shaping or conditioning. *Examples: single-point tools, including cemented carbide and coated carbide, tools—holders and inserts; drills, reamers, taps,*

threading dies, milling cutters, form tools, grinding wheels, saws, abrasive cloths and papers; punch and die sets; thread rolling dies, holes, and knurling tools.

Part-specific tools, restricted in application. These are tools designed, made, and/or bought to produce a specific part or assembly, or a family of parts and assemblies. *Examples: forging dies, diecast dies, and patterns; forming dies, compound dies, progressive dies, and slitting dies; jigs, positioners.*

The variety of these items is greater than the variety of parts that will be made using them.

10.2 DATA CAPTURE AND IDENTIFICATION

Gathering information for the kinds of items defined in Section 10.1 ranges from relatively simple to extremely difficult. The easier categories are machinery (and equipment) and nonperishable tooling.

Figure 10.2 shows a list of machine tools by category, including their capital equipment property number, machine working dimensions, and the like. The tool design shown in Figure 10.3 identifies the function of the tool and its identifier. The cost department can provide relevant cost data and life history, as well.

The remaining categories of data are not so easily identified. More often than not, the supporting data are scattered, inconsistent, and incomplete. Far too often, the only records available are the most recent year's purchases. Some idea of the problem can be appreciated by considering the various sources of data other than purchase orders that have been experienced. Of the following list, rarely will one data source suffice without the need for additional data research:

Flow sheets of a process
Maintenance records and repair manuals
Stock control cards
Bin cards, held where the items are stored
Traveling requisitions for the replenishment cycle

These are the most reliable sources of information short of examining each item individually.

Even so, a great deal of research is necessary to fully identify the majority of items. Especially is this true of items that are bought as

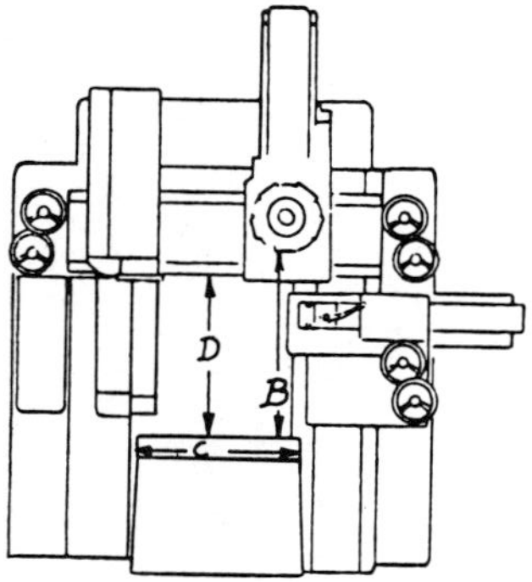

A - Maximum swing
B - Maximum height
C - Chuck diameter
D - Maximum height under rail
E - Facing height
F - Swivel of head
G - Horse power

VERTICAL BORING MILLS

No.	Name	Turret head	Side head	L.H.ram	R.H.ram	Thread, taper	Max. swing A	Max. height B	Chuck dia. C	Under rail D	Facing height E	Swivel of head F	Horse power G	Special Work	Year Mar.	Year Pur
9544	32" G & L	✓					42	40	32	28	-	30	40		59	59
9501	42" G & L	✓	✓				52	49	42	37	37	30	50	Navy & Imp.	58	58
9468	42" G & L	✓	✓				52	49	42	37	37	30	50	Navy & Imp.	58	58
9330	36" Bullard	✓	✓				45	41	36	30	21	30	30	Bearing Housings	56	56
0548	46" Bullard	✓	✓				60	51	46	40	46	30	40	Digital read out		75

Continued

6

Figure 10.2 Data capture document example for machinery and equipment.

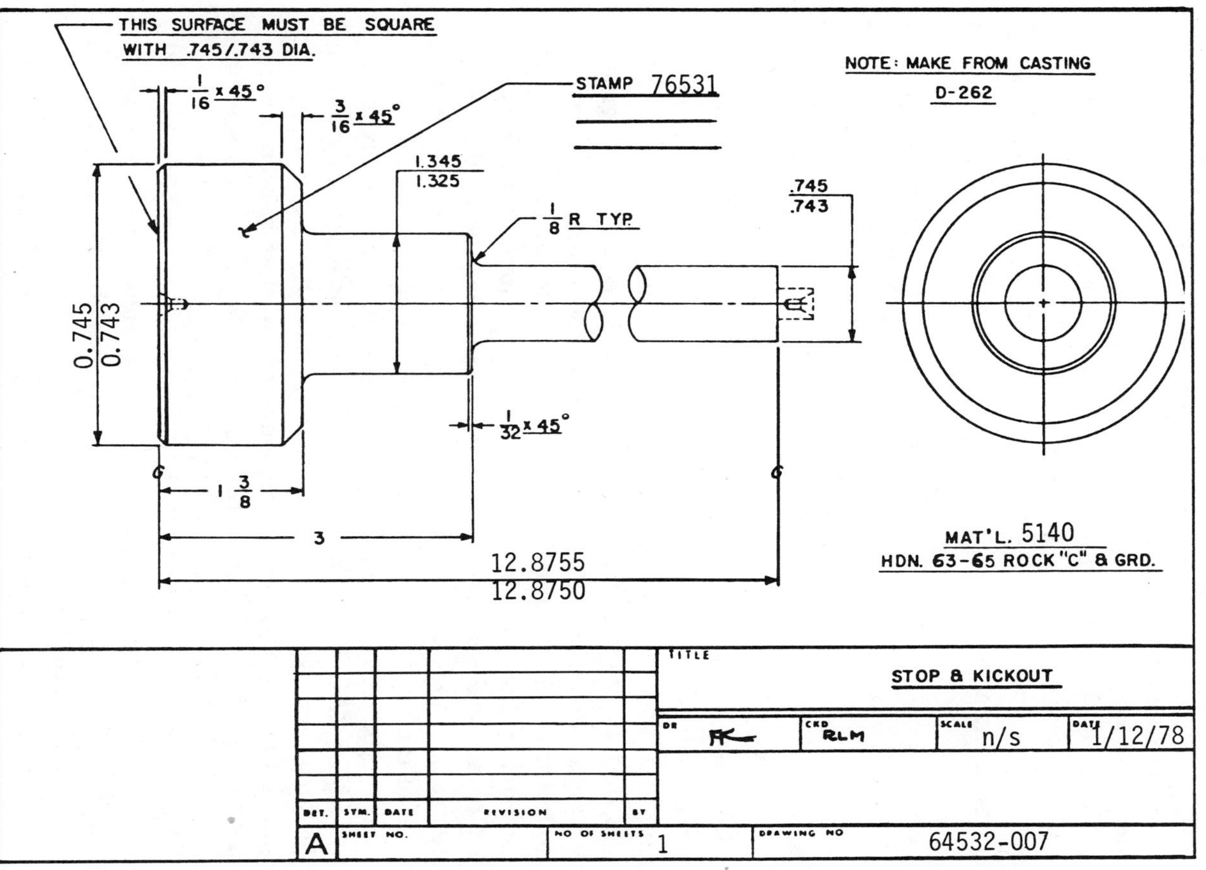

Figure 10.3 Data capture document for part-specific tool design.

makers' proprietary spare parts. Frequently, the numbering system used by the supplying machinery makers, in their internal system, is the cause of the problem.

Figure 10.4 is taken from the maintenance manual of a machine tool manufacturer. The items shown are keyed by detail number to a parts list, as shown in Figure 10.5. Examine, for example, items with detail numbers as follows:

316
326
327
329
331
332

The key number for this particular machine cross-references to the part number used by this firm to identify their parts. The part numbers, descriptions, and amount used are shown in Figure 10.5.

It does not take a great deal of study to deduce that, logically, these items are not proprietary spare parts exclusively designed and made for or by the machine maker. The pipe and/or tube fittings are immediately suspect as common, garden-variety items made to an industry or national standard. These same items may be used by many other machine tool manufacturers. In point of fact, these same items may be in the product that is manufactured using the machine tool.

As can be seen from Figure 10.5, the parameters necessary to fully identify what 4AC-316 (part number 89879) actually are not available. Under such circumstances, it is necessary to have the maker supply the missing information either by drawings or by clear written descriptions—that is, if the maker is disposed so to cooperate, and some are not.

Now, examine Figures 10.4 and 10.5 and locate part key 4AC-270. The description says that the item is a "plug screw." Is it a proprietary item or is it not? Obviously, with the data available here, it cannot be logically determined one way or the other.

The same is true for items shown in Figures 10.6 and 10.7. Examine 4ADV-227 and 236, both tapered roller bearings; cones and cups. In Figure 10.7 these are identified as "bearings-roller," with part numbers 171948 and 171945, respectively. As is obvious, these could be

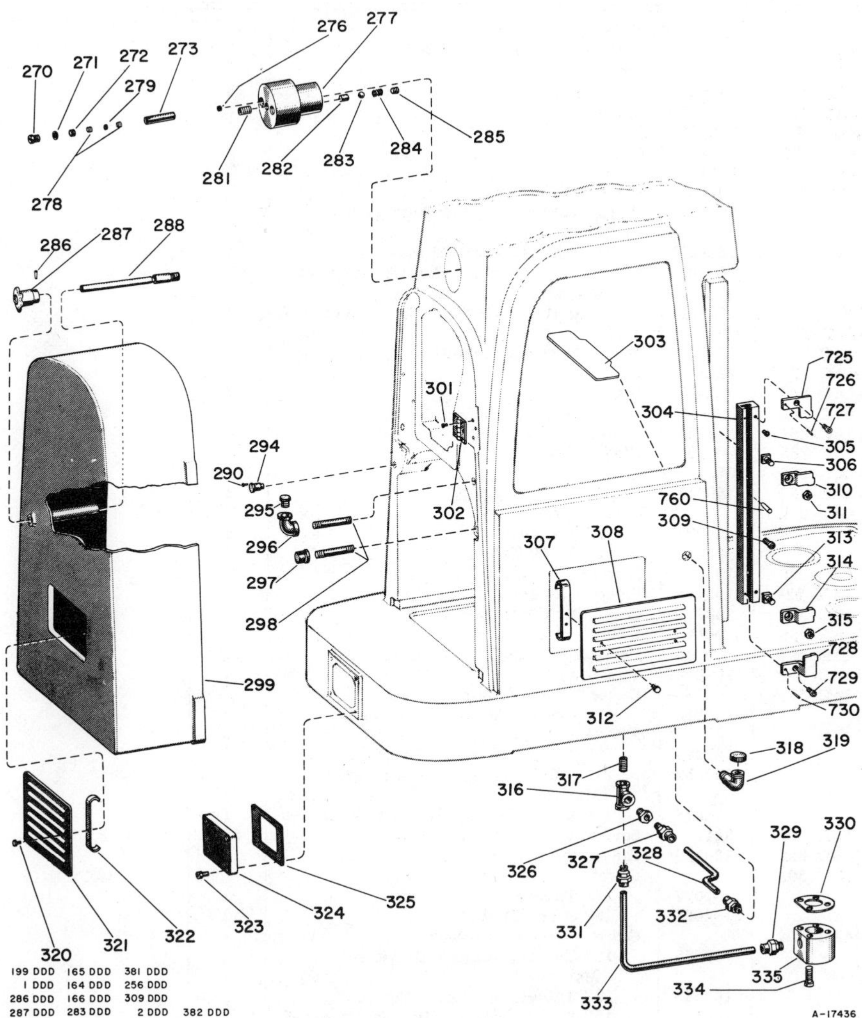

Figure 10.4 Exploded view of column dogs and covers in a service manual. (Courtesy of Cincinnati Milacron, Inc.)

KEY NO.	PART NO.	PART NAME	AMT. USED
4AC-270	76449	Plug - Screw	2
4AC-271	129871	Gasket	2
4AC-272	160155	Screw	2
4AC-273	159293	Case	2
4AC-276	2356	Screw	2
4AC-277		Body - Dial Regulator	
	160159	12" Tables, Plain & Universal Machines	1
	162949	12" Tables, Vertical Machines . . .	1
	160161	15" & 16" Tables, Plain & Universal Machines	1
	160162	15" & 16" Tables, Vertical Machines.	1
4AC-278	2356	Screw	4
4AC-279	167060	Disc - Dynamic Resistance	8
4AC-281	2248	Plug.	1
4AC-282	257472	Seat - Ball	2
4AC-283	30154	Ball - Steel	2
4AC-284	167558	Spring	2
4AC-285	76645	Plug - Pipe	2
4AC-286	663	Pin	1
4AC-287	153411	Knob	1
4AC-288		Stud - Knob	
	164005	12" Tables	1
	164004	15" & 16" Tables	1
4AC-290	1650	Screw	1
4AM-294	106991	Safety - Switch	1
4AC-295	3987	Cap - Oil Filler	1
4AC-296	12814	Elbow - 3/4".	1
4AC-297	200786	Cap - 3/4" Pipe	1
4AC-298	144624	Pipe - List	1
4AC-299		Cover - Rear Motor	
	147980	12" Tables, Plain & Universal Machines	1
	147979	12" Tables, Vertical Machines . . .	1
	147550	15" & 16" Tables, Plain & Universal Machines	1
	152678	15" & 16" Tables, Vertical Machines.	1
4AC-301	3218	Screw	12
4AC-302	128759	Hinge - Motor Cover	2
4AC-303		Cover - Oil Pocket	
	151077	12" Tables	1
	194021	15" & 16" Tables	1
4AC-304		Guide - Column Vertical	
	169585	12" Tables, Plain & Universal Machines	1
	169586	12" Tables, Vertical Machines . . .	1
	169587	15" & 16" Tables, Plain & Universal Machines	1
	169588	15" & 16" Tables, Vertical Machines.	1
4AC-305	2322	Screw	2
4AC-306	3361	Bolt	1
4AC-307		Spring - Cover	
	157281	12" Tables	2
	157280	15" & 16" Tables	2

Figure 10.5 Parts list for items shown in Figure 10.4. (Courtesy of Cincinnati Milacron, Inc.)

KEY NO.	PART NO.	PART NAME	AMT. USED
4AC-308		Cover - Motor Opening	
	143980	12" Tables	1
	146267	15" & 16" Tables	1
4AC-309		Screw	
	2322	Plain & Universal Machines	4
	2322	Vertical Machines	3
4AC-310		Dog - Upper Column	
	169591	12" Tables	1
	169595	15" & 16" Tables	1
4AC-311	2057	Nut	1
4AC-312	4173	Rivet	4
4AC-313	3361	Bolt	1
4AC-314		Dog - Lower Column	
	169592	12" Tables	1
	169597	15" & 16" Tables	1
4AC-315	2057	Nut	1
4AC-316	89879	Tee	1
4AC-317	24815	Nipple	1
4AC-318	3986	Cap	1
4AC-319	211167	Elbow - Oil Filler	1
4AC-320	4295	Rivet	4
4AC-321		Cover - Motor Opening	
	143980	12" Tables	1
	146267	15" & 16" Tables	1
4AC-322		Spring - Cover	
	157281	12" Tables	2
	157280	15" & 16" Tables	2
4AC-323	3403	Screw	4
4AC-324	100026	Cover - On Base Reservoir	1
4AC-325	219931	Gasket	1
4AC-326	89138	Bushing - Reducing.	1
4AC-327	79729	Fitting - Straight	1
4AC-328		Tubing - Oil Filler	
	151076	12" Tables	1
	147527	15" & 16" Tables	1
4AC-329	79730	Fitting - Straight	1
4AC-330	108729	Gasket - Sump Elevating Screw. . . .	1
4AC-331	79730	Fitting - Straight	1
4AC-332	79729	Fitting - Straight	1
4AC-333		Tubing - Sump	
	151075	12" Tables	1
	147510	15" & 16" Tables	1
4AC-334	3404	Screw	4
4AC-335	108725	Sump - Elevating Screw	1
4AC-725		Dog - Positive Stop, Upper	
	169590	12" Tables	1
	169594	15" & 16" Tables	1
4AC-726	424	Pin	1
4AC-727	124975	Screw	1
4AC-728		Dog - Positive Stop, Lower	
	169593	12" Tables	1
	169596	15" & 16" Tables	1
4AC-729	124975	Screw	1
4AC-730	424	Pin	1
4AC-760	663	Pin	1

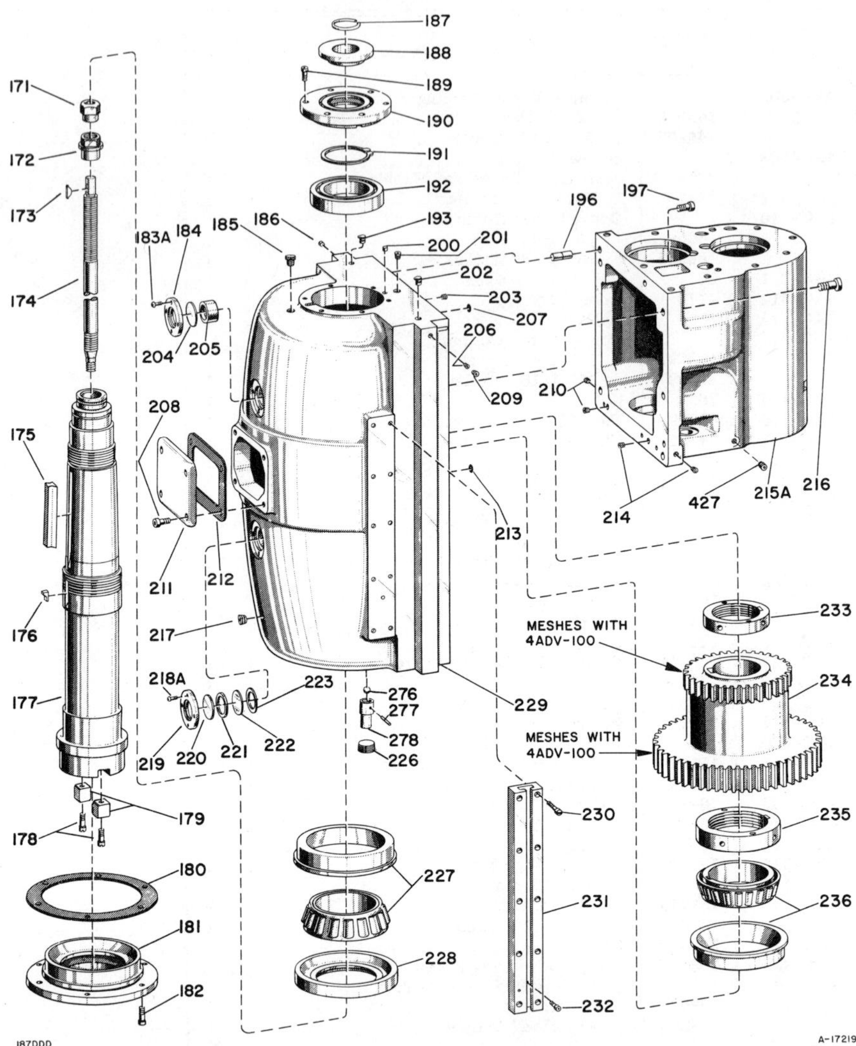

Figure 10.6 Exploded view of a vertical spindle shaft in a service manual. (Courtesy of Cincinnati Milacron, Inc.)

KEY NO.	PART NO.	PART NAME	AMT. USED
4ADV-171	3639	Bolt - Head	1
4ADV-172	3641	Nut	1
4ADV-173	3280	Key	1
4ADV-174	151643	Bolt - Arbor Draw In	1
4ADV-175	144834	Key - Square	1
4ADV-176	4004	Key - Lock Bearing	1
4ADV-177	171319	Spindle - Veftical	1
4ADV-178	1158	Screw	2
4ADV-179	3687	Key - Arbor Driving	2
4ADV-180	147456	Gasket - Bottom of Spindle	1
4ADV-181	147464	Retainer - Oil	1
4ADV-182	3399	Screw	6
4ADV-183A	2369	Screw - Button Head	4
4ADV-184	4075	Cover - Oil Gage	1
4ADV-185	64929	Plug - Pipe	1
4ADV-186	2245	Plug.	1
4ADV-187	210560	Ring - Retainer	1
4ADV-188	114882	Cap - Spindle.	1
4ADV-189	3399	Screw	6
4ADV-190	147458	Cap - Spindle.	1
4ADV-191	238199	Ring - Retaining.	1
4ADV-192	161372	Bearing - Roller	1
4ADV-193	83512	Oiler	1
4ADV-196	1430	Pin	2
4ADV-197	3343	Screw	1
4ADV-200	1962	Plug - Brass	1
4ADV-201	76645	Plug - Pipe	1
4ADV-202	83512	Oiler	1
4ADV-203	4236	Plug.	1
4ADV-204	3849	Disc - Oil Gage	1
4ADV-205	131986	Body - Oil Sight Gage	1
4ADV-206	4236	Plug.	1

KEY NO.	PART NO.	PART NAME	AMT. USED
4ADV-207	247090	Ring - "O"	2
4ADV-208	3399	Screw	4
4ADV-209	76644	Plug.	1
4ADV-210	76644	Plug.	2
4ADV-211	151631	Cover - Front	1
4ADV-212	151632	Gasket - Front Cover	1
4ADV-213	127468	Ring "O"	1
4ADV-214	76644	Plug - Pipe	2
4ADV-215A	184952	Housing - Transmission	1
4ADV-216	3408	Screw	6
4ADV-217	216607	Plug - Pipe	1
4ADV-218A	2369	Screw - Button Head	4
4ADV-219	4075	Cover - Oil Gage	1
4ADV-220	3849	Disc - Oil Gage	1
4ADV-221	4086	Gasket - Oil Gage	1
4ADV-222	60956	Dial - Oil Gage	1
4ADV-223	3850	Gasket - Oil Gage	1
4ADV-226	221981	Plug.	1
4ADV-227	171948	Bearing - Roller	1
4ADV-228	147463	Slinger - Oil	1
4ADV-229	150401	Body - Vertical Head	1
4ADV-230	3485	Screw	10
4ADV-231	144328	Bar - Tee Slot	1
4ADV-232	3396	Screw	1
4ADV-233	4275	Nut - Lock	1
4ADV-234	149623	Gear - On Spindle	1
4ADV-235	149843	Nut - Lock	1
4ADV-236	171945	Bearing - Roller	1
4ADV-276	30154	Ball - Steel	1
4ADV-277	3540	Pin	1
4ADV-278	165174	Check - Ball	1
4ADV-427	4236	Plug - Pipe	1

Figure 10.7 Parts list for items shown in Figure 10.6. (Courtesy of Cincinnati Milacron, Inc.)

special bearings or unrestricted-application bearings. On the same illustration, however, there can be little doubt that 4ADV-234 (part number 149623), "gear on spindle," is the maker's proprietary designed part. So the problem resolves itself to the following three sets of conditions:

1. If the spare part is probably a nonproprietary spare part, all that remains is to examine the box it is received and/or stored in for the maker's name and part number. That is, it will be other than the machinery maker's number.
2. If the part is very likely a proprietary part of the equipment maker's design, accept it as such, but query if this part is used on any other of the types of machines and/or models that may have been supplied by this maker.
3. If there is no documentation (i.e., part number other than maker's) and/or there is question as to whether it is an original-equipment manufacturer's (OEM) part, query the supplier for details to describe the item or items in question.

For future identification purposes, a system is established, together with a procedure to inspect all such items upon receipt. If a part number, either on the box or on the part, is different from that assigned by the machinery or equipment maker, it is to be notated and maintenance, purchasing, and variety control notified for query.

10.3 DEVELOPING THE CLASSIFICATION

Recall from Chapter 3 that there are four steps in the classification of an identified body of data: it is done in four steps:

1. Develop an analysis code based on firsthand experience with the data base. This usually results from a preliminary survey and a sampling of data.
2. Analyze the data by coding, using the analysis code.
3. Quantify the distribution.
4. Synthesize the classification to distribute the variety consistent with present and future goals.

Classification trees are developed as shown in Figure 10.8. This is a portion of a classification tree for pipe fittings. It is an n-polar tree, using an indent method of presenting the classification. It is done this way

PIPE FITTINGS

- Threaded
  - Alloy Steel (including stainless)
    - Adapters
    - Bushings
    - Couplings
    - Crosses
    - Elbows
    - Laterals
    - Nipples
    - Pigtail Syphons
    - Reducers
    - Reducing Connectors
    - Tees
    - Caps
    - Unions
    - Locknuts
    - Plugs
    - Union Fittings
  - Carbon Steel
    - Adapters
    - Bushings
    - Couplings
    - Crosses
    - Elbows
    - Laterals
    - Nipples
    - Pigtail Syphons
    - Reducers
    - Reducing Connectors
    - Caps
    - Plugs
    - Return Bends
    - Tees
    - Union Fittings
    - Unions
    - Locknuts
  - Iron
    - Adapters
    - Bushings
    - Caps
    - Couplings
    - Crosses
    - Elbows
    - Laterals
  - Iron (cont.)
    - Locknuts
    - Plugs
    - Reducers
    - Reducing Connectors
    - Return Bends
    - Tees
    - Union Fittings
    - Unions
  - Brass
    - Couplings
    - Nipples
    - Unions
    - Adaptors
    - Elbows
    - Bushings
    - Reducers
  - Bronze
    - Elbows
    - Bushings
    - Coupling
    - Reducer
    - Union
    - Tees
    - Crosses
    - Locknuts
    - Return Bends
    - Laterals
    - Caps
    - Plugs
  - Non-Metallic
    - Adapters
    - Bushings
    - Caps
    - Reducing Connectors
    - Elbows
    - Nipples
    - Plugs
    - Couplings
  - Weld
    - Iron
    - Carbon Steel
    - Stainless Steel
  - Flanged Fittings

Figure 10.8 Tree indent classification of MRO pipe fittings. (Copyright © Brisch, Birn & Partners, Inc.)

BASIC CLASSIFICATION		DEFINITION OF CLASSES		SUB-CLASSES			
MINING AND EXCAVATING	MINING AND EXCAVATING	MACHINES AND PLANT ITEMS. TOGETHER WITH THEIR SPECIFIC SPARES. INVOLVED DIRECTLY WITH THE EXTRACTION OF MINERALS.	1	1	DRILLING AND BORING		
				2	BLASTING		
				3	SUPPORTING AND TRAMMING		
				4	UNDERGROUND LOADERS MAKERS		A TO L
				5			M TO Z
				6	ORE CONTROL. SHAFT & HEADGEAR		
				7	SHAFT WINDERS, LIFTS		
				8	OPEN PIT EXCAVATORS		
				9	RESERVED		
	TRANSPORTING INCLUDING I.C. ENGINES	VEHICLES INCLUDING OFF HIGHWAY, ROAD, RAIL. AIR AND WATERBORNE USED FOR TRANSPORTING MINERALS AND PERSONNEL. TOGETHER WITH THEIR SPECIFIC PURPOSE SPARES. I.C ENGINES ARE ALSO INCLUDED.	2	1	OFF HIGHWAY VEHICLES	MAKERS	A TO L
				2			I TO P
				3			Q TO Z
				4	ROAD VEHICLES	EXCEPT COMMERCIAL	
				5		COMMERCIAL	
				6	RAIL, AIR, WATER TRANSPORT		
				7	I.C. ENGINES, GAS TURBINES		
				8	AUTO PROPRIETARY SPARES		
				9	RESERVED		
	CONVEYING AND CONDITIONING	MACHINES AND PLANT FOR THE CONVEYING OF SOLIDS. LIQUIDS AND GASES, TOGETHER WITH THEIR SPECIFIC PURPOSE SPARES. ALSO INCLUDED ARE AIR CONDITION ING AND DUST EXTRACTION PLANTS.	3	1	LIFT TRUCKS		
				2	CRANES		
				3	CONVEYORS		
				4	PUMPS	MAKERS	A TO L
				5			M TO Z
				6	COMPRESSORS VACUUM PUMPS		
				7	BLOWERS, FANS		
				8	VENTILATION, AIR & DUST PLANTS		
				9	RESERVED		
	MINERAL PROCESSING AND UTILITY SERVICES	MINERAL PROCESSING MACHINES AND PLANT, TOGETHER WITH THEIR SPECIFIC PURPOSE SPARES. ALSO INCLUDED ARE SPECIALISED AUXILIARY, BY-PRODUCT AND UTILITY PLANTS FOR COBALT, COAL, PYRITES, OXYGEN, ACID, STEAM, ELECTRICAL GENERATION, WATER AND SEWAGE DISPOSAL, LINERS ARE ALSO INCLUDED.	4	1	REDUCING AND CLASSIFYING		
				2	CONCENTRATING		
				3	LEACHING, ELECTROLYTIC REFINING		
				4	SMELTING, LINERS		
				5	REFINING O/T ELECTROLYTIC		
				6	SPECIALISED PLANTS	I:	
				7		II:	
				8	STEAM AND UTILITY PLANTS		
				9	RESERVED		

MAINTENANCE SUPPORT SERVICES	INSTRUMENTATION ELECTRONIC AND ELECTRICAL SUPPLIES	INSTRUMENTS, EXCEPT THOSE FOR MEASURING LENGTH, ANGLE OR OUTLINE – SEE 61500. ELECTRONIC AND ELECTRICAL SUPPLIES NORMALLY PURCHASED TO NATIONAL OR INTERNATIONAL STANDARDS AND / OR ACCEPTED TRADE DESCRIPTIONS, MAKERS REFERENCE NUMBERS.	5	1	INSTRUMENTATION	
				2	ELECTRONIC COMPONENTS	
				3	COMMUNICATION EQUIPMENT	
				4	STORAGE AND CONVERSION	
				5	ELECTRIC CABLES AND CORDS	
				6	O/H LINE, LIGHTING, HARDWARE	
				7	CIRCUIT CONTROL AND PROTECTION	
				8	MOTORS, GENERATORS	
				9	RESERVED	
	SERVICES EQUIPMENT AND SUPPLIES	SPECIFIED MAINTENANCE SERVICE EQUIPMENT AND SUPPLIES FOR THE SUPPORT OF THE MINING AND PROCESSING FUNCTIONS, AS SET OUT IN THE SUB-CLASS HEADINGS.	⑥	①	TOOLS, PORTABLE EQUIPMENT	
				2	DOMESTIC AND WELFARE EQUIPMENT	
				3	BUILDING SUPPLIES	
				4	CLOTHING, SAFETY, SIGNS	
				5	HYGIENE, FOOD, STOCK FEED	
				6	LABORATORY EQUIPMENT	
				7	MEDICAL AND SURGICAL SUPPLIES	
				8	OFFICE SUPPLIES	
				9	RESERVED	
	MECHANICAL HYDRAULIC AND PNEUMATIC COMPONENTS	ENGINEERING COMPONENTS NORMALLY PURCHASED TO NATIONAL OR INTERNATIONAL STANDARDS AND / OR ACCEPTED TRADE DESCRIPTIONS, MAKERS REFERENCE NUMBERS EXCLUDED FROM THIS CLASS ARE ITEMS ALREADY COVERED IN CLASSES 50000 and 60000.	7	1	FIXING AND	IMPERIAL SIZES
				2	FASTENINGS	METRIC SIZES
				3	VACANT	
				4	POWER TRANSMISSION	I:
				5	ELEMENTS	II:
				6	FLUID SEALS, PIPES AND FITTINGS	
				7	FLUID CONTROLLING	
				8	HYDRAULIC AND PNEUMATIC CONTROLS	
				9	RESERVED	
	PRIMARY MATERIALS	MANUFACTURING MATERIALS WHICH MAY BE SHAPELESS AND USED IN BULK (OIL, FUEL, CHEMICALS) OR REQUIRING AN OPERATION TO BE CARRIED OUT BEFORE THIS CAN BE USED. THIS OPERATION MAY BE AS SIMPLE AS CUTTING A PIECE FROM A BAR OR APPLYING PAINT WITH A BRUSH. REFRACTORY MATERIALS ARE ALSO INCLUDED, BUT NOT – **ELECTRIC CABLES AND CORDS** CLASS 50 000 – **PIPES, TUBES, HOSES WITH OR ADAPTED FOR END FITTINGS** CLASS 70 000	8	1	IRON AND CARBON STEEL	
				2	ALLOY AND TOOL STEEL	
				3	COPPER AND COPPER ALLOYS	
				4	METALS O/T 81 TO 83	
				5	NON-METALLICS EXCEPT 86 TO 88	
				6	CHEMICALS, FUELS, GASES	
				7	SPECIFIC PURPOSE MATERIALS	
				8	REFRACTORY MATERIALS	
				9	RESERVED	
			9			

Figure 10.9 MRO stores classification extracts from a Zambian copper mine.

SPANNERS, OPEN ENDED, CHROME VANADIUM				
Podger Ended, Cranked, King Dick Type, Bs. Hexagon And Whitworth Sizes				
7 /16 Ins. Bs., 3 /8 Ins. Whit.	61 452 2013 K		EACH	L
1 /2 Ins. Bs., 7 /16 Ins. Whit.	61 452 2014 L		EACH	M
9 /16 Ins. Bs., 1 /2 Ins. Whit.	61 452 2015 M		EACH	L
5 /8 Ins. Bs., 9 /16 Ins. Whit.	61 452 2016 N		EACH	M
11 /16 Ins. Bs., 5 /8 Ins. Whit.	61 452 2017 P		EACH	L
3 /4 Ins. Bs., 11 /16 Ins. Whit.	61 452 2018 Q		EACH	M
7 /8 Ins. Bs., 3 /4 Ins. Whit.	61 452 2019 R		EACH	L
1 Ins. Bs., 7 /8 Ins. Whit.	61 452 2021 U		EACH	LM
Double Open Ended, 15 Degree Angle Jaws, In Sets				
Set Of Six Spanners In Wallet, Sizes Range From 0 Ba To 11 Ba	61 452 2275 V		EACH	LM
Double Open Ended, Bs. Hexagon And Whitworth Sizes				
3 /16 Ins. X 1 /4 Ins. Bs., 1 /8 Ins. X 3 /16 Ins. Whit.	61 452 2311 J		EACH	CJ L
1 /4 Ins. X 5 /16 Ins. Bs., 3 /16 Ins. X 1 /4 Ins. Whit.	61 452 2312 K		EACH	J L
5 /16 Ins. X 3 /8 Ins. Bs., 1 /4 Ins. X 5 /16 Ins. Whit.	61 452 2313 L		EACH	CJ LM
3 /8 Ins. X 7 /16 Ins. Bs., 5 /16 Ins. X 3 /8 Ins. Whit.	61 452 2315 N		EACH	CJ LM
7 /16 Ins. X 1 /2 Ins. Bs., 3 /8 Ins. X 7 /16 Ins. Whit.	61 452 2316 P		EACH	CJ L
7 /16 Ins. X 9 /16 Ins. Bs., 3 /8 Ins. X 1 /2 Ins. Whit.	61 452 2317 Q		EACH	M

1 /2 Ins. X 9 /16 Ins. Bs., 7 /16 Ins. X 1 /2 Ins. Whit.	61 452 2318 R		EACH	C	L
1 /2 Ins. X 5 /8 Ins. Bs., 7 /16 Ins. X 9 /16 Ins. Whit.	61 452 2319 T		EACH		LM
9 /16 Ins. X 5 /8 Ins. Bs., 1 /2 Ins. X 9 /16 Ins. Whit.	61 452 2320 U		EACH		L
9 /16 Ins. X 11 /16 Ins. Bs., 1 /2 Ins. X 5 /8 Ins. Whit.	61 452 2321 V		EACH	C	LM
5 /8 Ins. X 11 /16 Ins. Bs., 9 /16 Ins. X 5 /8 Ins. Whit.	61 452 2322 W		EACH		L
5 /8 Ins. X 3 /4 Ins. Bs., 9 /16 Ins. X 11 /16 Ins. Whit.	61 452 2323 X		EACH	C	LM
11 /16 Ins. X 3 /4 Ins. Bs., 5 /8 Ins. X 11 /16 Ins. Whit.	61 452 2324 Y		EACH		L
11 /16 Ins. X 7 /8 Ins. Bs., 5 /8 Ins. X 3 /4 Ins. Whit.	61 452 2325 Z		EACH	J	
3 /4 Ins. X 7 /8 Ins. Bs., 11 /16 Ins. X 3 /4 Ins. Whit.	61 452 2326 A		EACH		LM
3 /4 Ins. X 15 /16 Ins. Bs., 11 /16 Ins. X 13 /16 Ins. Whit.	61 452 2327 B		EACH	J	L
7 /8 Ins. X 1 Ins. Bs., 3 /4 Ins. X 7 /8 Ins. Whit.	61 452 2328 C		EACH		LM
1 Ins. X 1.1 /8 Ins. Bs., 7 /8 Ins. X 1 Ins. Whit.	61 452 2329 D		EACH		L
1.1 /8 Ins. X 1.1 /4 Ins. Bs., 1 Ins. X 1.1 /8 Ins. Whit.	61 452 2331 F		EACH		M
Double Open Ended, King Dick Slim Series, Bs. Hexagon And Whitworth Sizes					
1 /4 Ins. X 5 /16 Ins. Bs., 3 /16 Ins. X 1 /4 Ins. Whit.	61 452 2372 A		EACH		M
5 /16 Ins. X 3 /8 Ins. Bs., 1 /4 Ins. X 5 /16 Ins. Whit.	61 452 2373 B		EACH		M
3 /8 Ins. X 7 /16 Ins. Bs., 5 /16 Ins. X 3 /8 Ins. Whit.	61 452 2374 C		EACH		M
7 /16 Ins. X 1 /2 Ins. Bs., 3 /8 Ins. X 7 /16 Ins. Whit.	61 452 2375 D		EACH		M
1 /2 Ins. X 9 /16 Ins. Bs., 7 /16 Ins. X 1 /2 Ins. Whit.	61 452 2377 F		EACH		M
9 /16 Ins. X 5 /8 Ins. Bs., 1 /2 Ins. X 9 /16 Ins. Whit.	61 452 2378 G		EACH		M

Figure 10.9 (continued)

solely to keep the presentation of information size consistent. It is the equivalent of a branching tree.

At this point it is proper to ask if the designation "in-product" and/or MRO are mutually exclusive terms. Obviously, they are not. Items that are classified elsewhere will be found in the classification that bears the classification-based code number.

The decision to classify and code MRO items before in-product materials and components has occurred a number of times in our experience. In chemical manufacturing firms; copper, coal, gold, diamond, and uranium mining; oil refineries; and food-processing industries, with which we have had condiderable experience, it is only logical. One firm uses wellhead gas and crude oil hydrocarbon feedstocks to manufacture thousands of different products. They support this process with nearly 200,000 MRO items, spending $140,000,000 annually while maintaining a $70,000,000 inventory.

Even manufacturing firms that produce an engineered end product may give priority to these kinds of items over in-product items. One such firm, about to spend a considerable amount on new plant facilities, felt that before equipping the plant and provisioning spare parts, they should know what was already available and/or where used. Another firm, which mass produces high-ticket products, wanted to reduce downtime on their lines and isolated machinery spare parts availability as a key factor in realizing their objective. Another firm found that they were paying as much as 14% more across the board for common-use items bought at five different locations. In each case, two facts stood out: (1) each had multiple locations, and (2) none had data visibility and/or control of the inventory of such items.

When the populations are consolidated and all items identified, classified, and coded, many desirable things can be done to reduce cost and/or improve profitability. This will be explained later in the chapter.

The resulting classification and family code may appear as shown in Figure 10.9. This is a classification and family code for consumable MRO items in the copper mining industry in Zambia. This code was developed for the federal consortium (NCCM) that resulted from government purchase of the interests of the four major producing firms. These codes apply to established mining sites and can accept new mines as they are developed in less civilized areas. In that respect, human support items to create and maintain a city in the wilderness with access mainly by air and lead times of 120 days or more for even simple spare parts make for a rather complex planning equation.

10.4 INTERFACE OF CLASSIFICATION-BASED CODE NUMBERS FOR PLANNING IN MANUFACTURING

That all the data items used in a manufacturing facility should be retrievable upon demand can hardly be disputed. Like hot dogs and apple pie, what could be more all-American from the manager's viewpoint? In Figures 10.10 and 10.11, the interface of various different populations of items are shown. The interplay, one with the other, for planning for manufacturing reveals the dependency upon data being visible and/or retrievable.

Figure 10.10 shows the end product or model number as key to the product bill of materials. The product is comprised of proprietary assemblies, commodities (nonproprietary parts and assemblies), proprietary piece parts, and raw materials.

Figure 10.11 shows in broad terms how to interrelate forecasted expected volume, factored by yield multiples, to account for scrap and mysterious disappearance of raw materials and commercial commodity items. All these items must be purchased presuming that inventory balances are properly managed and in control. The flow chart does not include a statement covering this vital activity (inventory of items quantities on hand), as it is presumed to be understood. There are other purchasing responsibilities (i.e., tools, proprietary parts, and assemblies bought out for any of several reasons).

Figure 10.11 also shows the factors and steps necessary to plan labor requirements. The box marked with an asterisk contains references to:

Cost-center codes	Class 0
Standard operation codes	Class 0
Standard times to perform the operation	Class 0
Machine codes	Class 7
Tooling codes, perishable and nonperishable	Class 6

The human resources, and the work they do, bear the Class 0 designation. Class 6 and 7 items relate to the assistance by machine tools, equipment, and tooling to produce the end product. The flow chart(s) shows, in general terms, the means to plan equipment needs and/or the capital budget. In one segment of the chart, the emphasis was on labor-hours; in another segment, the emphasis is on machine-hours. The same codes, however, provide the data common to both kinds of calculations. To many of you, this system is already functioning in the firms in which

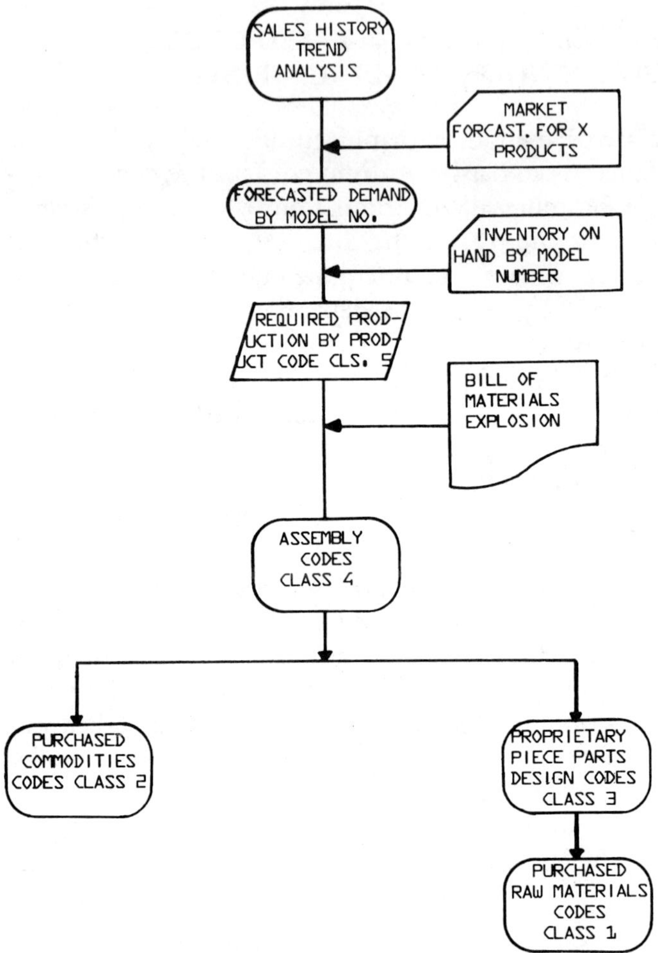

Figure 10.10 Flow chart of code numbers interface.

you manage. However, it has been our experience that the overview of codes interaction is missing.

Figure 10.11 is useful for pricing and cost-control purposes, as it shows how the product costs are accumulated at the customers' locations.

This interface of codes to plan objectives becomes more appreciated when viewed in this light. For this reason, you can see that if downtime factor is high—say, 20%—when equipment is bought at an

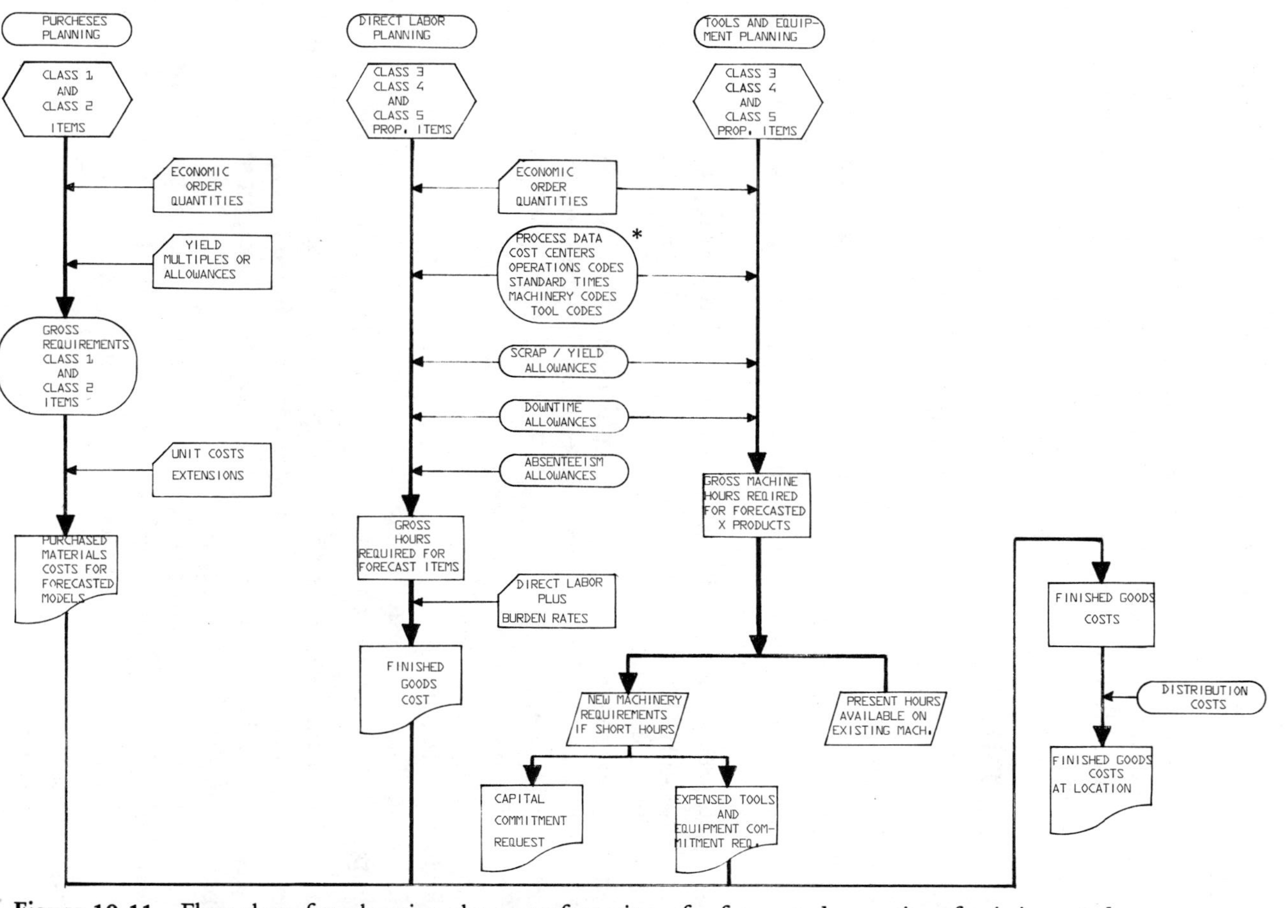

Figure 10.11 Flow chart for planning the manufacturing of a forecasted quantity of existing products.

expected 85% operational, 15% down, the effect is either overtime at inflated costs or lost throughput.

So to have control of the entire process necessary to manufacture a product, deliver it on time, and at a planned-for cost/sales price margin requires that all data relating to cost to do this be visible and/or retrievable.

10.5 BENEFITS

Benefits occur from many sources, as follows:

Increased yields or throughput
Streamlined purchasing function to take advantage of corporate usage, rather than individual procurements of common items at multiple sites (frequently, this means annual contracts with releases against the annual contract)
Reduced procurement transaction costs (fewer individual purchases)
Reduced unit costs
Balancing of inventories
Shifting equipment and/or spares from division to division, as the need arises and usage dictates, rather than buying new equipment and/or spares
Shifting tooling for the same reasons
Improving stock rotation
Reducing stock levels
Reducing the cost to carry inventory
Improved maintenance costs

How maintenance costs can be improved is a subject of interest to every manager. Everyone recognizes that maintenance expense is a necessary evil. Few managers, however, have done much more than put patches on the overall problem.

The experts in maintenance management have attempted to find ways and means to plan, organize, measure, and control labor. Few, however, have looked at all facets of maintenance costs, including:

Management
Administration
Materials
Labor
Yields

Some years ago, this problem was addressed in its totality by the writer and his organization. The result was Master Maintenance Controls. Standard data for time to inspect, diagnose, remove, repair, and replace were determined, as well as for construction, housekeeping, and sanitation. The all-embracing overview enabled standard times to be applied based on the identified and coded items affected in the maintenance operation.

The procedure is too detailed to explain in a portion of a chapter of a book on classification, coding, and group technology. It requires a book in its own right to do it justice. But the procedure does make it possible to improve greatly the cost of maintenance and improve the worth of the physical plant and yield in the bargain. The benefits listed are just a few of those that can be realized from the visibility of other than in-product items of overhead costs.

11

Management's Assessment of Classification, Coding, and Group Technology

11.0 WHAT MANAGEMENT HAS TO SAY

In the final analysis, only those who have had experience with the management tools described in this book can make fair comment. The case histories presented in this chapter are, in actuality, papers presented by users. These have been edited to remove references to consulting firms or to systems that are proprietarily identified with consulting firms. This is not intended as a slight to the consulting profession, of which this writer is a proud member. Rather, it is to focus attention on the use of the tool as a means to help management to manage better and not to sell the services of a consulting firm.

The papers reflect the impact of classification upon the functions for which the speakers are responsible within their respective firms. The controller speaks from his viewpoint, the standards controller from his, and so on.

However, two papers do reflect more of an overview of the total impact upon the specimen firms. These papers show the omnibus nature of the tools.

Although savings figures are mentioned, more often than not they are expressed in percentages and/or in conservative terms. In one of the case papers, however, there is supporting published documentation extracted from a government agency report. Although not part of that paper, it has been appended to the relevant firm's case.

11.1 CASE STUDY: PERKINS ENGINES LTD., PETERBOROUGH, ENGLAND, A SUBSIDIARY OF MASSEY-FERGUSON LTD.*

The Standard Controller's Requirements for Information Retrieval and Dissemination†

Introduction

I have been asked to talk on a paper entitled, "The Standard Controller's Requirements for Information Retrieval and Dissemination." It is to cover my company's reasons for the requirement of a logical classification and the acceptance of the coding system. The initial problems of installing the system, the day to day running of the system, and finally what benefits have we derived from it.

I have therefore decided that perhaps a brief "History of Perkins Engines," as a company, and its numbering systems prior to coding would set the background to our reasoning.

History of Perkins Engines

Perkins Engines started as a manufacturing unit in 1932, in a small workshop just off the centre of Peterborough, to manufacture high speed small dimensional diesel engines suitable for a variety of purposes, agricultural, automotive and marine with reputedly 3 men and a boy. Frank Perkins, the Founder, being one of the men.

It grew steadily up to 1939, but with the war it really took off and became a major supplier of engines for the war effort. Soon after the war the original premises at Queen Street were felt to be bursting at the seams and a new factory was built at Eastfield on the outskirts of Peterborough. Perkins moved the main factory and offices to Eastfield around 1948/49. This has been expanded time and again adding new workshops and transfer lines until the company now has about 10,000 employees at Peterborough alone; there are also manufacturing units world wide either wholly owned by Perkins or manufacturing under License.

The company is quoted as being as being the largest producer of small diesel engines in the world, capable of building 1500 units a day, Consisting of a variety of types from 3 cyl. to V8's.

In 1959 the company was taken over and became part of the Massey-Ferguson organisation, producing engines specifically for agricultural purposes

*Classification and coding commenced in 1956; installed in 1957.

†Presented by J. G. Moult, Standards Controller, Perkins Engines.

for M.F., but also continuing with old established customers in other applications, some in competition with M.F. with agricultural implements, and so we have continued up to the present time. That gentlemen is a very brief history of Perkins Engines.

Part Numbering Systems

Primary Materials

We had no type of part number for Primary materials prior to coding and draughtsmen and designers had more or less a free hand to select a material or size of material without too much opposition. Materials were known only by their BS Number 970 EN 1A, 1449 EN 2B, etc. Also in those early days pre-1957 certain materials were in short supply and alternatives were agreed and used, the surplus going back into stock. As Design Dept. did not know what was in stock the surplus materials were inclined to be left on the shelves unused and gradually they built up to a hefty stock-holding. I'll come back to this subject a little later.

Commodities

Our method of numbering proprietary parts was to allocate a block of numbers within a 5 digit system to each supplier in alphabetical order, i.e. C.A.V. 30,000, Clayton Dewandre 30,101, etc.

We soon ran out of the 100 numbers given to each supplier and had to allocate a new series some distance from the original, thereby splitting a manufacturers numerical sequence and mixing filters with pumps, dynamos with starters, etc.

Perkins Designed Components/Assemblies

Perkins had two part numbering systems prior to coding:

First a 6 digit number was used. This system was based on a block of numbers being allocated to each engine type, and all major components being given a number within that block i.e. our six cylinder engine of those days was known as the P6, and was allocated the series 280000. The second two digits denoted the component part, 28, 25, 00 indicating a cyl. head. The last two were used as 01-09 for variations of engine build lists, and 01-99 for variations of the component. As with commodities we ran out of numbers.

As the range of engines increased and the variety of components with the possibility of interchange between engines grew, the same piston, valves, springs, etc., the numbering system lost quite a lot of its original meaning.

A new numbering system was introduced in the late 1940's comprising seven digits. This was an alphabetical system with Badges, Bearings, Belts starting at 001, 004 and 007 respectively through to Washers at 092. The first 3 digits being significant, the remaining 4 being issued numerically from a register. Other titles were added as an afterthought, throwing this system out of sequence.

Groups of these numbers, notably Flywheels, Brackets, Shafts, Flywheel Housings, soon began to expand into the hundreds. Brackets we had 1000 + , Flywheels 400 + , Manifolds 300 + and so on.

In "Brackets" of course we had small pressed steel right angled items mixed in with cast engine mounting brackets, bearing no similarity to one another at all.

Retrieval relied upon memory or a long search through full size drawings, it was easier to draw a new one, and this is what happened, creating even more components to add to the ever growing variety. A new part number was issued on demand from a register by one of the clerical staff in the Drawing Stores area.

One of the most disturbing factors was that all flywheels, for instance, are very similar in basic shape, and as all but very few of our flywheels were made of C.I., a new piece of pattern equipment had to be laid down for every new flywheel, when it was quite feasible that a pattern already existed, from which the new wheel could be easily be machined. This problem of course aplied to everything, Housings, Sumps, Manifolds, etc.

Another problem was the designer's interpretation of the title; is it a bracket or a support, is it a link or a lever, dependent on the title of the drawing was the area from which a number was allocated, therefore brackets and supports, similar in shape and application would be at opposite ends of the numbering system. This then was the reason for another change.

Around about 1956 it was decided therefore that something much more sophisticated was required, and the company in its wisdom decided on the hierarchical method of classification.

Once the decision had been made to use the coding system the practicabilities had to be faced.

1. The work was to be done in those days at the consultant's offices in Victoria Street, London.
2. A representative of Perkins had to attend four days a week to liaise with their coding and classifying personnel. This was to agree on behalf of Perkins the best way to code and classify and to answer all questions with regard to the product whether of technical or practical nature. This meant staying in a hotel from Monday to Thursday, returning to the office on Friday to clarify all queries.
3. 15,000 reduced prints, which at that time was about ⅓ of our total, and representing a good cross section of our component parts, had to be prepared first by micro-filming and then printing 8 × 5 size in black and white for ease of handling and subsequent filing in the coding reference file. This job was too large for Perkins facilities at that time so a camera and

operator was hired from Kodak Ltd. who with two of Perkins staff completed the job in five weeks of filming and printing. These were then despatched to London and we were ready to start with the coding.

Perkins had agreed to use 4 classes.

1. Primary materials
2. Commodities and proprietary parts
3. Perkins designed components
4. Subassemblies

The first three running simultaneously, followed by Class 4 Sub-Assy's.

What Our Requirements Were

The Engineering Design Division of any company requires a functional system of classification for the materials it uses, the commodities it buys from outside, and for its own designed component and assembly drawings, to enable it to work efficiently and smoothly with a minimum amount of retrieval and investigating time by engineers, designers and draughtsmen.

Classification is only a means, a tool for designers, but its impact on any industrial organisation is manifold. Some of the most important decisions taken in any company outside the board room are the ones taken on the drawing board. The design featues of a component, its dimensions and the material from which it is to be made, are determined in the Engineering division, and its repercussions are felt far outside it.

Purchasing, Planning, Production, Stores, Cost and Spares supply are all affected. It is therefore of the utmost importance that the Drawing Office have ready access to all information that may influence design, and avoid duplication.

The first function of a coding and classification system is to provide just that.

Logic of Classification

A classification by its nature is an orderly, systematic structure.

A design of a classificatory structure must satisfy two basic requirements: (i) Comprehensiveness, and (ii) mutual exclusiveness of its categories. In other words, the scope has first to be wide enough to bring in all of the items that need to be included in the various classes, and second, the definition of the classes must be exact enough to ensure that there is only one place for every item, and that place is the same for every user of the classification.

The underlying logic is simple: every question must have a unique and unambiguous binary answer: "Yes" or "No"; "True" or "False"; "Present" or "Absent"; "Included" or "Excluded"; and so on.

Now this type of classification is tailor made to suit a company's individual requirements and it is therefore necessary for each company to indicate where a large concentration of certain types of items are to be found, so that provision within the classification can be made.

We at Perkins for instance have a large number of pipes and pipe assemblies. It was therefore considered necessary to have one complete group for each of these items, with flywheels, we required 3 complete sub-groups, two for pulleys, two for gears and so on, tailor-made to our requirements.

It took me 14 months to complete the coding and classification to our mutual satisfaction and by August 1957 we were ready to start with the new numbering system.

With the completion of the classification and coding, we then had to start to make the best use of it.

Within Classes 1 and 2 we were introduced to a system of Preferred or Non-Preferred items. The selection of items for the Preferred or Non-Preferred status depended upon B. Standards, Manufacturing Standards, availability, cost, quantities required, etc., the final word resting with Perkins Engines.

What Our First Achievements Were From the Coding

Primary Materials

During our classification process we had, with mat. lab., agreed standard specifications of material as preferred stock, and listed our existing stock within those specifications by allocating a code number.

All stock not included within the standard specification, was either used up on known application or put out for sale.

The stated saving in both cash from sales and storage space was in excess of £100,000. We now had a control area, no new drawing would be issued without a material code number, no new code number would be issued without an investigation by the coding officer; he had the authority to question the reason and refer back to supervisor or mat. lab. for confirmation. Hence designers would now look into the code book and select from an existing range. The Preferred and Non-Preferred ranges were indicated by coloured sheets, "green" for Preferred and "pink" for Non-Preferred.

Commodities

Commodities were a complex problem ranging from Standard hardware to Starter Motors, Compressors, Fuel Injection Pumps, Cold Starting Devices, Filters, etc.

Very often interchangeable items, possibly to British or Manufacturers' Standards, can be purchased from different suppliers—Oil Seals, Injectors, Filters, etc.

In the past, as stated, these would have been listed under the supplier in alphabetical order. Now they are all listed in their appropriate area of the coding system, i.e. 2100, 2300 General Hardware, 2400 for Jointing, Sealing, Pipe Fittings, Cooling and Compressors, 2500 and 2600 for Transmission Elements, 2700 for Circlips, Springs, Grommets and Recording Instruments, 2800 Electrical Items.

Within these General headings are Sub-divisions relating still further to each item's individual title and range of specifications or sizes.

Retrieval then becomes a matter of minutes.

Components

Components as stated were numbered from an alphabetical system, but we require a functional system of classification and we feel that we've got this from this classification system.

Within the Class 3 classification, we have eight simple groups.

As a manufacturing company one thing was missing. We had no foundry or forging facilities, and it was necessary to identify a casting/forging produced elsewhere, for instance in Birmingham or Sheffield and then being moved to another destination for machining.

Our problem is, that except in very rare cases we do not have a separate casting drawing and so the casting suppliers work to the fully dimensioned machining drawing. We therefore stipulated that all items machined from castings/forgings would need to have a separate identification, we agreed that in a block of 10 numbers allocated (slide) the zero would always indicate the casting number and all components machined from it would follow in sequence, i.e. 37716210 casting, machined components 37716–211,212 etc. This system was therefore included into our classification.

It did mean that instead of having 999 components in one series of numbers, i.e. 37716–001 through 37716–999, we only had 99 where castings were concerned.

Because of our requirements in this situation we were able to adjust the coding to incorporate all the extra space required.

We have now modified this slightly by adding a stroke number i.e. 37716–210/3 to indicate the master drawing from which the latest pattern is made; this absorbs any modifications to the pattern.

Sub-Assemblies are defined as being of two or more parts which can be taken apart without being destroyed.

This posed a problem as we have pipe assemblies with fixed olive connections on the ends but loose union nuts in between which cannot be removed. These it was agreed would go into Class 3, but the same assembly with a removable clip would go into Class 4, a borderline case, but it satisfied our method of engine build listing and as long as the rules are clearly defined and adhered to, all is well.

A similar rule applies to fabrications, if all the separate items are welded, brazed or fixed by some other permanent means, then it becomes a single piece part in Class 3.

What Our Initial Problems Were and How Were They Solved

Technical problems, re: materials, commodities and components, were overcome as they arose during the classification, by arranging to have meetings with the various Perkins departments; Friday or Monday being the most satisfactory, and as all departments readily co-operated, we had few problems.

Practical problems were the main headache. How were we to collate all the information necessary to enable the consultants to do their job?

As far as materials were concerned, it was a matter of listing all materials in the steel stores, general stores and paint and oil stores, and from that massive list delete all items not considered as production requirement.

As it turned out all the steel stores and paint/oil stores items were included, but General Stores had to be separated—maintenance from production, all our stores records were on Kalamazoo sheets so we photographed every sheet and went through them item by item crossing out those not required. They were then given the marked up sheets and we handed our headache to them.

With Classes 2 and 3 we had separate detail drawings or stock lists of hardware. As previously stated about 15,000 items were involved. All of these items carried either a 5 or 7 digit number and were filmed and printed by Kodak.

It had already been decided that Perkins items would retain their existing part number, the reason being that too much confusion would result by altering the part number of all engine component parts, as they were quoted on countless pieces of documentation not only at Peterborough but World Wide.

This meant that all the 8 × 5 prints had to be stamped with the new type code number. This was done with a numbering machine by hand. Concurrently with this, a Perkins to coding number cross ref. was compiled.

The 8 × 5 prints were put into packs and the whole system filed into cabinets of the correct size, this became known as the coding reference file.

When it became necessary to alter a drawing, it also became necessary for the consultants to check to see if the alteration in any way changed the part number. This didn't happen very often because if the change was so great as to warrant this, it was usually a requirement of the alteration to issue a new part number for identification purposes.

It did mean however that a new 8 × 5 print was required for the file and the code number had to be located in the cross ref. and then stamped on the new print.

To rectify this we gradually had the Code Ref. No. transferred on to the original drawing by tracers working from a cross ref.

Whilst the work was going along at London, we were also becoming organised at Peterborough and all new and altered drawings were automatically routed to the micro-film and printing area, and an 8 × 5 print produced. Also all current items not sent to London were being systematically printed. All of these were filed in Perkins part number order and the new and altered replacements made to keep the file up to date.

These were to be coded and added to the data file at a later date.

All new drawings from the 1st August 1957 were given code No's. As numbers were no longer taken out from a register because of the variety of family numbers within the Coding System, we decided to use a grid, and this has proved very successful and easily maintained with a minimum amount of recording to do.

Finally, without the co-operation of the people you wish to use the system, you may as well not have bothered, so a number of lectures to Designers, Purchase, Manufacturing Depts. including senior storemen and foremen were held to familiarise the company with our new toy. We mesmerised them with its simplicity and the great majority became converts.

(A) 1. WHAT THE SYSTEM IS DESIGNED TO DO AND WHAT BENEFITS WE ARE GETTING

1.1 As a functional System, it classifies into preferred and non-preferred items, the

Primary materials the company uses
Commodities or proprietary parts it buys
Components and assemblies of Company design

This enables the Company to work efficiently with a minimum of retrieval and investigation time spent by Draughtsmen, Designers, Engineers, Purchasing Staff and Plant Supervision in obtaining information on existing standards, materials and component designs.

The four classes in current use in Perkins are:

Class 1—Primary materials
Class 2—Commodities or proprietary parts
Class 3—Perkins own design parts
Class 4—Perkins own design sub-assemblies and assemblies

1.2 Provide a facility whereby rationalisation and standardisation can be achieved. This covers such aspects as

Material specification
Size range of hardware
Variety of product components and assemblies

1.3 Control the addition to existing variety in all these Classes, and to elminate the possibility of duplication. This has a direct contribution to the maintenance of minimum component and manufacturing costs, storage, and spares requirements.

(B) 1. *WHAT BENEFITS DOES IT PROVIDE?*
THE SYSTEM

1.1 Manuals are provided for Classes 1 and 2 which are distributed throughout the Company, particularly in the Design Areas for immediate reference with respect to available materials and components which are considered of standard type. Less than 1% addition has been made in Class 1 to ranges established originally on installation of the system in 1957. Such is the control the System provides on Primary Materials in Class 1. This survey was carried out in 1974. Since that time additions have practically halted. The primary materials manual has been revised to change specs in accordance with BS.970, 1449, etc.

1.2 It provides a basic reference facility, particularly for Design Staff, of all Perkins design components and assemblies in a data bank form code Ref. File for efficient retrieval.

1.3 It ensures that the content of the file is organised in such a way that it can be virtually guaranteed that there is no existing item when new components are submitted and accepted for classification or coding.

1.4 With the facilities available, the result is to reduce the number of new parts designed by the adoption of existing designs or by the modification of existing designs to accommodate a new requirement.

(B) 2. *THE PAST POLICY OF STANDARDISATION*

2.1 In coding Classes 1 and 2, Preferred items are established by Supplier/B.S./Volume criteria.

In coding Classes 3 and 4, there were no Standards laid down, but recently we have raised our own Semi-Standards and Composite Drawings and identify Preferred items in these two types of drawings. These are selected by Volume/Design/Sales/Cost criteria and will be regularly reviewed. These provide a positive control facility in Perkins own design parts Classes.

2.2 The coding system has been successful in challenging new or all-but-the-same components, within the scope defined above. It is significant that no new part or material can be given a number without being screened by the coding Control. In coding Classes 3 and 4, there has been an estimated successful challenge of 24% on an average of 2600 new production component part numbers issued per year.

These successes are made up of (A) eliminating 2 new parts designed per week which need not have been drawn if further thought had been given. (B) Eliminating existing parts by incorporating existing Parts within new parts designed. (C) Routine retrieval of existing designs from coding file to satisfy new specifications.

A similar success rate is achieved on Prototype drawings which are produced at an average of 2500 drawings per year.

In coding Class 2, the figure is around 10% on 200 new items per year. In Classes 1 and 2 standard Preferred items are of the type which can be strictly controlled.

(C) 1. HOW ARE THE BENEFITS TO BE ACHIEVED? CURRENT USE OF THE SYSTEM

1.1 By the constant application of disciplines and controls provided by the System itself. These controls are manipulated by a group of Standards Staff working in co-operation with Design Staff and Engineers, and are strictly adhered to, ensuring there is no duplication or near duplication of parts.

(C) 2. DEVELOPMENT OF THE SYSTEM, APPLIED TO ENGINEERING COMPONENTS—CURRENT DEVELOPMENTS

2.1 To develop the coding System facility by preparing Tabulations *by families of parts* classified by shape within Class 3 and by assembly type within Class 4. These tabulations to be included in the data file to improve the retrieval facility.

2.2 Prepare ratonalised *Composite drawings of major component types* to reduce the number of drawings needed for a design, and provide a basic facility that will produce subsequent standardisation.

2.3 Produce *Semi-Standards* which set up a range of components of own design within coding Class 3 established by the Company. The Semi-Standard is set up on a Company Standard basis, based on our knowledge of design requirements for range of size, cost, design suitability, manufacturing suitability, volume of use, and specification usage.

The above developments are supplementary to the coding System and improve its efficiency, allowing the benefits of efficient retrieval, rationalisation and standardisation to be achieved.

These developments are already being working on, and have been for the past 6 years. Benefits are acknowledged by Engineering and Supply Divisions as extremely positive.

(C) 3. DEVELOPMENT OF THE SYSTEM, APPLIED TO ENGINEERING COMPONENTS—PROPOSED DEVELOPMENTS

3.1 The day-to-day operation of coding is a routine function which takes place in the maintenance of existing file content and its manipulation regarding availability of existing materials and designs, and in its application to the classification and coding

and new requirements. The benefit is difficult to quantify, but it is positive in all these respects.

Improvements in retrieval time by applying Computer Techniques could be achieved. Visual display could be provided of existing components in the data coding file nearest to the design/specification required by the Design Staff. Such data as the current volume, cost and manufacturing processes could be provided, which is normally retrieved at present or not readily availably to Engineering.

3.2 The degree to which the coding System influences the Company operation depends entirely upon the authority given to the Standards Controller, who is responsible for its functioning.

3.3 Adequate Staffing of the right level of experience and ability will provide an ongoing facility to thoroughly vet new designs and to rationalise existing variety to establish standards which are not inflexible but which meet the Company's need for the Company product at any point in time.

11.2 CASE STUDY: CINCINNATI MILACRON, INC.*

Simplified and Standardised Engineering Design for Manufacturing Cost Reduction†

I am pleased to be here and to tell you of our experience with classification and coding. My prepared paper retraces steps you have already heard from other speakers so I will dispense with these and try to cover those points which, I feel, are unique to Cincinnati Milacron.

Our interest in classification and coding goes back some 11 or 12 years. We had given a great deal of thought to this subject. We investigated and experimented with two commercially available universal systems during this period. Also, one of our people developed a coding system on his own. The two universal systems proved unsatisfactory for our needs. This doesn't mean that other firms might not benefit from them, but they simply didn't provide the kind of benefits that we had decided must be available to us at Milacron.

In 1975, just about 10 years after we first became interested in classification and coding, we decided to try to find the answer for our specific problems of data retrieval once again. The impetus came from our manufacturing engineers. They had been experimenting—or should I say trying to justify using one of the proprietary universal systems. But after 18 months with little or no success in trying to use the system as a basis of computer-aided generative

*Installed 1976.

†Presented by H. K. Brown, Chief Engineer, Cincinnati Milacron USA.

process planning and computer-aided industrial engineering, conceded that there were too many faults with the system and on advice from our management abandoned it.

They, the manufacturing engineers, were still convinced that classification and coding of engineering and manufacturing data was an imperative for the future of the firm. We in engineering were in agreement so we decided to form a committee of all functional activities in the firm which could benefit from having engineering data and information visible and easily retrievable.

We asked each functionary, i.e. department head, to develop a work list. The list was to include all of the transactions that each made where visibility was essential to the efficient carrying out of their respective missions and the decisions taken on the merit of visible relevant factual data.

We re-evaluated all the universal systems—including the previous two that we had experience with, plus, this time, we included consulting people. They were asked to examine the documents listing the needs and wants of the overall organisation and present their ideas.

The tailored approach, i.e. developing a unique classification that catered for our specific needs and reflecting our specific data appeared to be the best solution to our needs. Together, we agreed that the best place to start was our proprietary piece part designs. We decided for three reasons

1. We had (or thought we had) very good guidance for most of our purchased raw materials.
2. We have very comprehensive standards for many of our purchased commercial mechanical items—bearings, fasteners and the like.
3. We had considerable pressure from our manufacturing engineers to attack this class of items which is their primary problem area.

In our firm, the funding of projects such as this one has to compete with other projects. To gain management approval requires that:

1. Our project must make a return of at least 25/30% after taxes annually.
2. It must stand high in a competitive ranking of all other projects under consideration.
3. There must be credibility for the documentation supporting the request.

We prepared our project request using the gold mine principle. This simply means examining the mother lode of benefits to determine the maximum amount of ore contained and then estimating at what rate you can mine it.

We estimated an annual savings figure and projected what return-on-investment would satisfy our board criteria. We then determined the expected rate using the cost figures for the project as proposed by our consultants. The rate-of-return using their cost projections exceeded the board guidelines by a substantial margin. We applied a factor to reduce the rate to account for any unforeseen difficulties—of which some developed. The figure we claimed was conservative (the very minimum we could expect) as our projects are audited periodically to verify costs and savings claimed. The project ranked well with its competitors and was approved.

Almost from the beginning of implementation (about 7½ months after the development began) the retrieval rate of useable existing designed parts was much higher than we had dared hope. In the machine tool industry, there is conderable customizing of our standard machines. Customer options and specialised fixturing make for a constant flow of new parts from our 245 designers and draftsmen.

Much to our surprise, we found that not only were simple parts being retrieved, but complex parts as well (and in about the same proportion).

Slide 1 This is a simple part of which we have hundreds. It is a round or cylindrical part with a threaded through going centre hole (single O.D.—single I.D.) with a threaded set screw hole at right angles to the centre hole.

Slide 2 This part is more complex. It is a button stop of hardened steel that is round, with multiple diameters and the major diameter has a spherical end.

Slide 3 This part is quite complex. Made from a casting, it has six finished surfaces, compound angles and a close tolerance on surface relationships.

Slide 4 This part, a casting, is even more complex, with holes, cam surfaces and the like. The interesting thing about this part is that its original application was for use on a milling machine. This retrieval was for use on an injection moulding machine. These are two quite diverse machines.

Our average rate of retrieval for the first 18 months of implementation has been 33%. That is, 33% of the time, our designers can find an existing design and use it *as is* in the new application. Another 25% of the time, a design can be found that by modification can be used in a new application. This may be a "make-from" type of modification or a logical extension of the configuration for a larger part or a smaller part. In either of these cases, much of the design work has been done, and the down stream manufacturing engineers and tool designers avoid unnecessary work.

Slide 5 Typical of the "make-from" assistance given designers is this casting, machined all surfaces. The projecting end was sawed off—everything else remained the same.

Slides 6, 7 and 8 [These] show other related examples where we were able to virtually eliminate any drafting, even though new parts numbers had to be created.

Slide 9 [This] shows a report that is computer-generated. It is a cross-reference in our present part number order—referencing CIMCODE family number. There is a second version of the same report by file inversion that is in CIMCODE order.

The remaining columns list cost data, among other things, and configuration level when a part number has had an engineering change made upon it. Similar parts in a family—that is to say those that are closely similar yet are different, are so indicated.

We had an interesting use of this report by our tool designers. They had tried in the past without success, to build multiple part tools—tools that could be used to make two or more parts. With this listing, candidate parts are immediately visible to them. They have also been designing tools so that they have adjustable features in anticipation of additions to the family, in addition to multiple part tools.

We are getting considerable numbers of inquiries from our manufacturing engineers. While they use the listing, they also want to see the other family members to find which is most closely similar to the one they have to process.

By finding likely candidates, they are able to modify the process record sheet to accommodate the differences without disturbing the operations that are the same. This has caused us to consider setting up files of the reduced size (11" × 8") drawings not only in design groups but in functions that are not physically located near us.

Currently, we have one man servicing our 245 people in engineering. There are two others trained as backstops, but only one operating the system. That's not quite right because all new hires for our department spend a minimum of two months in CIMCODE functional operation.

Two things, both beneficial, happen to new employees because of this.

A. They are indoctrinated and oriented as to our design practices, and are familiarised with the kind of parts we design and make.
B. They learn what the system can do for them and are sold its capability as an information retrieval tool.

The latter is important to us because the system is purely voluntary. Each new man added to our staff is an enthusiastic user of the system. And word does get around. More and more, day by day, our older employees commence to use the system. Every time a new part that passes through the system is found to have a useable design already in the files, because the designer did not use the system, the drawing is torn in two and handed back to the designer.

This only happens once. However reluctant these designers may have been in the beginning, they never fail to use the system after that.

We have kept this whole project low key, using persuasion and example rather than dogma. At Milacron, we believe this to be the wise course. We are proud of the history of our firm and the fine engineers and designers who have helped make it what it is. Our system has been designed to make continued use of these fine people by being able to use, wherever relevant and feasible, the best they had to offer. Some have long since passed on but we constantly build upon their expertise. They all thus continue to contribute even though they are no longer physically there. This is possible because what they did is immediately visible today.

In the early months of the project, we made an unusual and quite valuable contribution to management. Our *New Products Research* Group indicated that perhaps there was a need for a center grinding machine without automatic gauging. Their reasoning was that in developing nations especially, there must be a need for a machine where a grind might be called for but where tolerances were looser than .0005. After a suitable period of discussion, we volunteered to examine our CIMCODE files and see how many of our own parts calling for a cylindrical grind had tolerance greater than .0005. Out of more than 1200 such parts, we found only one. The project was cancelled immediately. It took one man about four hours to examine the families where a grind would be likely to appear in the processing of the parts. We prevented a considerable expenditure for designing, prototyping and promoting an unsaleable new product.

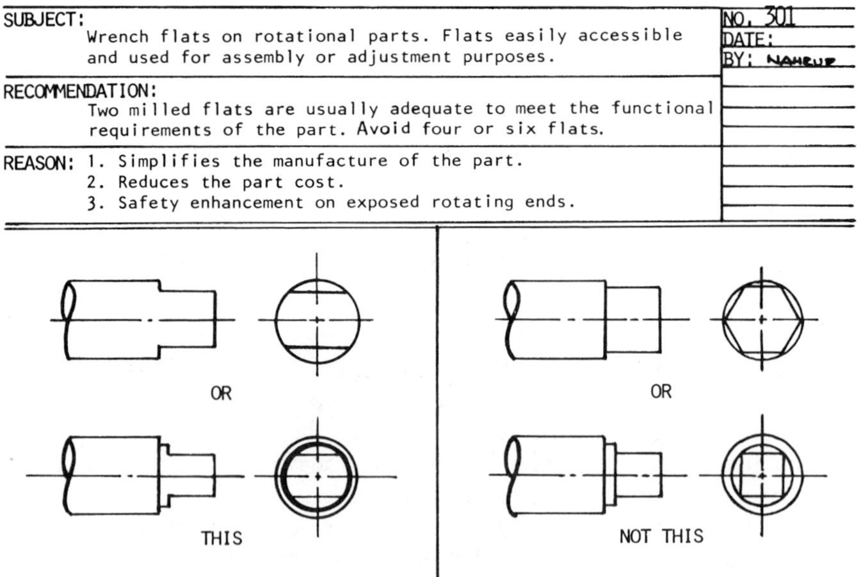
SUBJECT: Wrench flats on rotational parts. Flats easily accessible and used for assembly or adjustment purposes.

NO. 301
DATE:
BY:

RECOMMENDATION: Two milled flats are usually adequate to meet the functional requirements of the part. Avoid four or six flats.

REASON: 1. Simplifies the manufacture of the part.
2. Reduces the part cost.
3. Safety enhancement on exposed rotating ends.

Figure 11.1 Work simplification tips for designers, No. 301.

As I said earlier on, we have our projects audited by our internal auditing group. Our first audit revealed that the entire programme cost was returned in just under 3 months, a 400% plus rate of return per annum. This included the cost of consultancy, our internal salaries, the cost of photocopying our drawings, clerical costs—the lot. Benefits not quantified are represented by the following slides:

[Figure 11.1] is a Work Simplification Tip produced by our Engineering Standards Group. Where a wrenching portion is needed on a part, two parallel flats are recommended rather than a square or hex (as was the prior practice). This greatly reduces manufacturing costs.

Another is for hole patterns for round flat parts with holes on the bolt circle (cover plates).

Because drafting machines make it *convenient* to design four or six hole patterns—that is what our designers did. The Work Simplification Tip suggests 3 hole patterns for round parts up to the size shown and 5 hole patterns above that size. It costs us $4 each hole!

Square and rectangular covers will have four rather than six or eight held patterns. The cost reduction is considerable.

When possible, end mill created keyways are to be avoided in favour of straight tooth milled or woodruff keyways. [See Figure 11.2.]

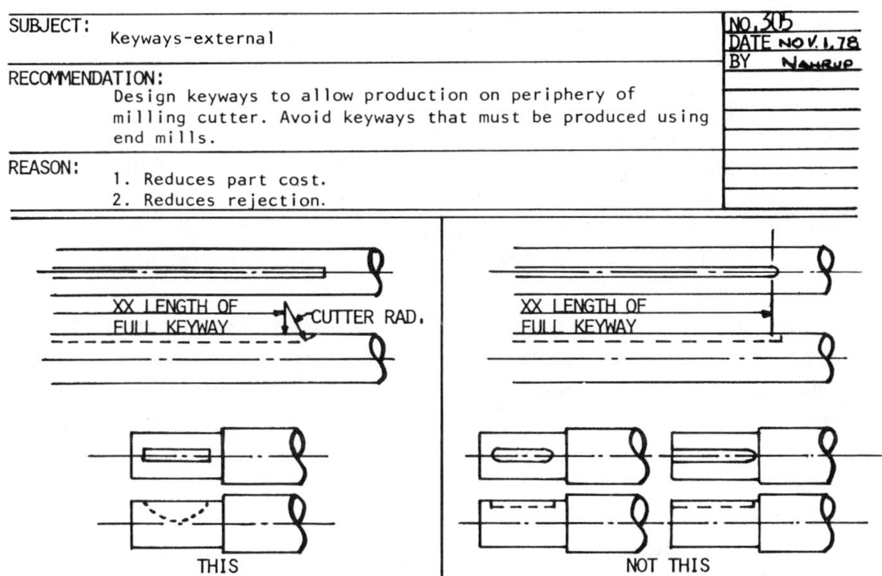

Figure 11.2 Work simplification tips for designers, No. 305.

These are only three of more than one hundred such Works Simplification Tips. They are inserted in the family files so that the designers see them prior to designing the part.

In conclusion our CIMCODE system has more than satisfied our manifold demands. The list of transactions which were prepared by the affected functionals have been satisfied. The design retrieval rate of 33% design prevention and 25% design assistance exceeded our estimates. The return-on-investment figure has impressed our management so that we have approval to commence step two of our long range plan—classification and coding of our commerical items. CIMCODE has also made it possible to proceed with our plans to computer-aid our designing activities. We are currently involved in adding to that capability now. But most important to us has been the improvement in the communications between our designers and our manufacturing people. CIMCODE is the common link which makes data visible to both functions and our new design standards mean that we are designing manufacturing cost reduction into our products. It has increased, not only design productivity, but has improved the effectiveness of everyone in the chain from concept of design to the manufacture of the part.

11.3 CASE STUDY: HOOVER WORLDWIDE COMPANY*

The Impact of Classification and Coding on Manufacturing Systems and Procedures†

As you no doubt will have noticed from your programme outline, the subject of this session is the impact of classification and coding on manufacturing systems. Obviously my session will be based very much towards my own company, but I believe the principles still apply. I do not intend to cover the intricacies of systematically arranging similar items into suitably selected categories, this has or will be handled, no doubt, by eminently more qualified people.

First of all I think it might be helpful if I introduced myself and my Company properly.

My name is Pat Tuhey and I am the project leader of manufacturing systems development at Hoover Ltd. Hoover are a medium to large sized organisation, employing some 25,000 people in the international organisation in order to market and where appropriate manufacture a variety of electrical appliances—vacuum cleaners, washing machines, tumble and spin dryers, refrigerators, freezers, irons, fractional horsepower motors and toasters—to name but a few.

The Company being a worldwide organisation—our products are sold by

*Installed 1970.

†Presented by P. Tuhey, Systems Department, Hoover Limited.

some 60,000 dealers in five Continents—is broadly speaking divided into two distinct regions. Eastern and Western. We in the U.K., which I will concentrate on, are part of the Eastern Region and generally speaking are the Design Centre responsible for all white goods, by that I mean washing machines, refrigerators and freezers as opposed to our floorcare products or irons.

Within the region are five major company production or manufacturing centres. Cambuslang—our Scottish factory based in Lanarkshire, Merthyr Tydfil in Glamorgan, Wales, Dijon in France, Meadowbank in Australia and Perivale in Middlesex. Perivale is the headquarters of the Eastern Region Hoover Organisation, as well as being the centre of the British Hoover Engineering and Design. In addition of course we have overseas licencees and agents.

Each of these production centres have drawing offices, and there our problem started! Drawings permeate all of the business, they are the basic identification and specification of the materials, parts and assemblies used in all products, not only Hoover ones. Around 1969 our drawing number system ran into trouble; over the years it had become overloaded, outdated and basically inefficient. The numbers themselves had no significance, they were no help in identifying unknown parts or avoiding design duplication. In most cases an individual's memory was the only aid to retrieval. Each drawing office issued numbers sequentially. Our system was a number system—it serialised or numbered a drawing. What was on the drawing was anyone's guess!

As a result of all these factors—not the least being that we were rapidly running out of numbers—Classification and Coding consultants were asked to survey the whole situation, and present their findings and appropriate recommendations to Executive Management. As a result of that report the decision was made to install the Hoover Worldwide Classification and Coding System throughout the Company.

This was not a difficult or indeed revolutionary decision for the need for classification systems had long been recognised within the organisation, for the complexity of modern industry requires good and sound methods of controlling the large number of items that are involved in the manufacture and servicing of products—whether our products or not.

The new C/C system would cover six basic categories.

1. Commodites not to Hoover design
2. Single piece parts (components) to Hoover design
3. Sub-assemblies and sub-arrangements to Hoover design
4. Descriptive items and packaging
5. Primary materials
6. Products (models) including accessories

Other categories such as tools, plant and equipment, etc. would be tackled at a later date. The Hoover code would be an 8 digit number with 5–3 notation.

Before installing the system, a comprehensive training programme was entered into for all managerial and technical grades, explaining the necessity for the system and what we as a Company expect to get from it. At the same time the precise mechanics or procedures for operating the system were agreed and they too were presented to a wide spectrum of Company personnel. The programme's objectives were to underline, to all levels of personnel within the Company, the potential benefits to be derived from intelligent use of the system and, secondly, to explain how the system would operate. Before all that occurred, however, a lot of work had been undertaken by consultant and Hoover personnel establishing the techniques and procedures to be used within Hoover by the classification and coding people. This entailed a detailed examination of some 80,000 drawings to enable the classification code books to be established. It is these code books that allow the systematic arrangement of similar items into suitably selected categories.

By this time, coding offices had been established at all three U.K. plants where all drawings would be classified and then uniquely coded. Drawings of like items in the form of A5 size prints were grouped together in pockets, a separate pocket being established for each series (or family) number. In the case of commodities and primary materials, the items were listed on computer printout, or as we call them, a raw materials catalogue and a commodity catalogue. This approach has enabled us to eliminate drawings of these two classes almost completely. The classification and coding function was undertaken by classification and coding officers, situated at our satelite locations. This has now changed so that all codes are issued from our central H.Q. at Perivale. This enables us to be more consistent in our approach to the system, and it ensures that design and production engineers do, as a matter of course, check existing designs and specifications when developing new ones.

That, in general terms, was how we came to introduce the Hoover Worldwide Classification and Coding system.

The impact of the system on the Company as a whole, however, was at once fascinating and obviously quite profitable.

Now, in order to understand how the system affected Hoover Limited, I would like to briefly mention those departments and activities where the impact was most significant.

Design Engineering Department

Remembering that our old numbering system had become outdated, overloaded and inefficient, the numbers having no significance, being no help in identifying unknown parts or helping in variety reduction or avoidance of design duplication. In short, the numbers at that time were a tag number! Bearing all that in mind then, it is obviously not going to surprise anyone that the impact was immediate!! In addition to the main benefits of classification and coding like variety reduction and control, standardisation and rationalisation, immediate improvements were apparent within the drawing system as a whole. For example, "used-on" references were deleted from drawing and a mechanised parts list (BOM) system was implemented which incorporated easy company wide "where used" information for the first time. Table drawings were implemented which has reduced drafting time. Numbers themselves were used more efficiently. A good example of this was to be found in the practice of allocating a drawing number simply to print on the drawing "same as whatever." Needless to say this practice has now ceased.

The big benefit however has undoubtedly been in retrieving information as an aid to rationalisation and standardisation exercises. Our new numbering system can be deciphered for intelligence, to describe shape, function and other basic features. Our current system of storing like items in pockets enables our designers to make use of existing designs when developing new ones. This quick reference to the file holdings is also a valuable aid in ensuring that numbers are not allocated to a new item if a suitable one is already in existence. For since the average cost within Hoover, to introduce a new piece part, is currently around £5,500, you will appreciate the potential savings that are possible with a system of this sort.

We estimate that improved information retrieval techniques, of which the system is an important part, have improved the operating efficiency of our drawing office and associated sections by something in the order of 3–7%. In cash terms, it is quite substantial for the Eastern region. (Many other savings are difficult to measure.)

Production Engineering Department

Have for the first time ever complete control of purchased materials. Production Engineers, like the Design Engineers, have immediate access via visual display units to current materials and their costs and detailed specifications, when deciding on precise material and tooling needs.

Other departments actively using the system include production control and purchasing departments. Their interest often relates to possible alternative items, when out of stock situations occur. In those cases where substitutes are identified a variety reduction exercise will follow. The system also is an aid to Purchasing department when comparing supplier prices on bought-out items.

Our Pricing and Estimating departments are continually examining parts with different costs but with the same family or series number. Where there are no good reasons for this situation, they will via Purchasing Department and Line Management challenge the supplier or department concerned.

The Stores area now accept that the coding system goes a long way to alleviating the age-old problem of wrongly identifying parts. A sample component is classified and numbered enabling paper-work to be corrected where appropriate. This assistance is greatly used during annual stock-takes and those instances where goods from our suppliers have been delivered with no visible identification.

Marketing and Sales Division also use the system greatly. We introduced a complete products catalogue which enabled them to ascertain quickly just how many divergencies of a particular product we were manufacturing and the reasons for those divergencies. I must say here, that, what was found was quite startling and so a detailed examination occurred which enabled us to delete a significant number of those products. The reason this had not occurred before was simple—to retrieve information appertaining to products was too time consuming, it was easier to introduce a new model divergency!!

In addition, of course, now *all* of our employees can broadly speaking identify a type of product from the numbers.

I would like to briefly mention, in my capacity as a manufacturing systems man, another perhaps hidden benefit of the system, and that is in the way in which the system has, indirectly perhaps, made us as a Company look at the way our other systems work, and the way in which we were, and are, organised to support them. Let me elaborate. We decided at an early stage that we would use the Class code number as the identifying number of all items within the system. That decision, ultimately, was to be instrumental in highlighting just how autonomous the various systems like costing, materials and production control, etc. had become. Indeed, I think it is not exaggeration to say that, in certain circumstances, these divisional systems did not relate to each other causing extreme difficulty when implementing Company policy decisions. For example, we found part numbers were being raised by our Materials department to cater for items going on outside relief, when really the correct procedure should have been to use operation numbers. These numbers were obviously not used, or indeed even recognised by other departments. I leave you to imagine the sort of confusion that emerged from that sort of situation.

Another good example of an autonomous system wending merrily away was our service system. In this important area of Company activity, a completely separate numbering system had been established—a service pack numbering system. We quickly investigated the need for such a system, decided it contravened the basic principle of 'one number—one thing' and therefore insisted that Service division participate within the Company numbering system. One final example—we found that within the three U.K. plants, seven separate numbers existed for Fullers Earth. This meant seven PRR's, seven purchase orders, seven separate receiving reports, seven separate bin cards in stores for the same thing. We were almost tempted to enquire how the devil our supplier coped.

These sorts of situations, and many others, finally convinced Hoover Limited that manufacturing systems in particular should not be allowed to develop as autonomous segments. They should be integrated Company systems, hence our decision to utilise data system techniques as opposed to the old method of developing specific information systems for specific people in specific departments at specific points in time.

I have no doubt in my own mind that besides the tangible savings related to variety reduction and control, standardisation and rationalisation, utilisation of redundant stock, identification of unknown items, information retrieval and improvements to the Company's drawing systems, the new classification and coding system was indirectly responsible for a strategy for manufacturing systems development being established within Hoover Limited, including the organisation and procedures required to support that strategy.

Finally then, we at Hoover have now been using the Hoover Worldwide Classification and Coding System for some 8 years within the six classes I have mentioned:

Commodities
Single piece parts
Assemblies
Descriptive items
Raw materials
Products

and I, as a manufacturing systems man, look forward to the day when we extend the system to cater for tools, plant and equipment. For then, I believe, we can confidently say that we have an efficient and worthwhile Management Information and Retrieval System.

11.4 CASE STUDY: COLES CRANES LTD., SUNDERLAND, ENGLAND, A SUBSIDIARY OF ACROW LTD.*

New Areas for Cost and Productivity Improvement through Simplification and Standardisation†

Foreword

In presenting this paper I am going to deal with four basic aspects.

First, why did Coles, a long established company, find it necessary to adopt a classification and coding system.

Second, how did we classify and code our parts. Here I will try to illustrate the methods used.

*Classification and coding commenced in 1976; installed in 1977.
†Presented by C. Bunting, Technical Services Manager, Coles Cranes Ltd.

Third, what type of problems did we encounter in this task.

Fourth, a summary of the benefits accrued from the implementation of Cee-Code.

Why Classify and Code?

We at Coles have experienced remarkable growth during the past thirty years. We have acquired three other crane manufacturing plants with their own design activities. We have expanded and diversified our product lines, and we have sophisticated our products. Naturally with such activity our management problems have also grown. Many of our systems and procedures, while viable through some period of our long history, have now become inadequate to effectively control our operation. The challenges of the future have dictated that we organise ourselves with new capabilities to meet the challenge.

Exceeding our growth has been the growth of the computer industry—both in the capabilities of computing hardware and the sophistication of computing software. Twenty years ago, our computers addressed accounting programs almost exclusively, but industry has learned that in any circumstance demanding the manipulation of any great mass of detail, including engineering and manufacturing data, computers are the most effective way to handle the manipulation. However, to make the maximum use of the computers available to us today, there are several minimum requirements.

The single most important requirement is a scheme of identification of our componentry and materials by a numbering system which will serve as a record address in data storage—a system of numbering which will be meaningful to all of the users of the data, while allowing the computer the maximum options in data manipulation.

We have long recognised the need for more capability in manipulating data relating to our in-product componentry and materials. The mass of data pertinent to these has reached staggering proportions. The problem we have faced was the lack of any data element in our parts records which could provide the means whereby useful manipulation could be done to provide the needed visibility to make effective management possible.

But before undertaking the development of a new scheme (Cee-Code), we examined the possibility of using our present part numbers. Unfortunately, our scheme for numbering parts does not describe the part in any way—it is simply a sequentially assigned, non-significant number. They are of no value in working the problem of data visibility which is so badly needed.

All of our parts have a name, so we examined the possibility of using the part name to handle and retrieve data but, again, our parts names are not reliable because parts tend to be named based on the application of the parts—not a viable means for data manipulation, as you will see. In determining whether a new part should go into our product, part names were found to be useless.

For example, if a new "Support" is thought to be required, it would literally be impossible to examine every other existing "Support" to determine whether one of them would serve the new purpose. The name "Support" can mean many different things and there are many, many parts with the words "Support" in their names.

While many components, such as supports, are entirely different but have the same name, it is equally true that many very similar components have different names.

If components which are similar in function or shape can be examined as a "Family" of parts, then many important things become possible (more about this later).

Our conclusion: Neither part names nor our present part numbering scheme was a viable means of working the data handling problem which we so badly needed.

We have developed Cee-Code to accomplish some very significant things on its own, and to provide a springboard to solutions to many other vexing problems. It is a parts and materials numbering system, based upon classification.

I will give an explanation of the principles later, but for now, it is important to understand that classification is a system of logic whereby like things are brought together by virtue of their similarities until "Families" of like components are established.

Possibly the best known classification is the Linnaean Taxonomic classification of all living things. It (and all other good classifications) are based upon four basic principles. These are:

1. The logic of the classification must be mutually exclusive. That is to say it must allow one and only one place for any specific item.
2. The classification logic parameters must be based on the permanent characteristics of the items being classified.
3. The classification must be all embracing. That is to say it must include provision for all existing and future items.
4. The classification must be tailored to suit the total needs of the host company and the parameters of classification must be consistent for these total needs.

We examined the Universe, which is our business and we found our total production effort could be divided into several distinct and mutually exclusive Sub-Universes, or classes.

Currently Cee-Code includes Classes 1, 2, 3 and 4:

1. Raw materials
2. Proprietary components
3. Our designed components classified by shape
4. Our designed components classified by function

We determined that these areas represent our most pressing visibility problems.

However, we recognise that we have other needs and therefore we have structured our overall Cee-Code concept to cater for those needs in the future.

How We Classified and Coded

As mentioned previously the main objective of the Cee-Code classification and coding system is to provide

1. Data visibility and retrieval
2. Logical numbering system

As the first step, raw materials and componentry were divided into four classes.

Class 1. Raw materials
Class 2. Parts from supplier catalogues, classified by function
Class 3. Single piece parts designed by Coles, classified by shape
Class 4. Sub assemblies, designed by Coles, classified by function

The classification is binary. This means that each level in the classification hierarchy is made mutually exclusive by a simple "yes, it is" or "no, it is not" decision.

The classification hierarchy is married to a code using a copyrighted matrix format method developed by our consultants.

These illustrate how a "family code" number is assigned. To identify an item uniquely, a "given name" called a sector code is assigned. (Our Cee-Code is seven digits in length).

The computer file address is the Cee-Code. All relevant technical, use, cost, and management data relating to a given Cee-Code are retrievable by accessing the computer file by the Cee-Code.

This illustrates one method of data display. While we can interact with the computer at a visual display terminal and see the same data, we distribute hard copy for users who do not or should not use a terminal.

On this catalogue display are such data as is necessary for selection and controlling/managing. This includes the following:

1. Material description, size, and other technical data
2. Status for future design and/or disposition
3. Using locations (i.e., various factories)

Types of Problems Encountered

Now that sounds fairly simple and straightforward. However, as you are no doubt aware nothing is quite that simple so here I will attempt to show the problems we encountered in marrying this coding and classification system to our requirements.

As might be expected our first problem was strategic in that in retrospect we would have approached the classification problem in a more complete manner. Basically we ignored Rules 3 and 4 mentioned before.

Memo

3. The classification must be all embracing. That is to say it must include provision for all existing and future items.
4. The classification must be tailored to suit the total needs of the host company and the parameters of classification must be consistent for these total needs.

We in fact classified Classes 1 and 2 in isolation and succumbed to the temptation to assign marginal cases to Classes 3 and 4 to be dealt with later. Consquently when we came to do Classes 3 and 4 we were faced with a revision requirement to some of our parameters of Class 2 and thus we were continually back tracking to keep our classification and coding consistent. Therefore this emphasises that in adopting this type of system, careful attention must be paid to the basic rules. Secondly, rather than a problem with the system, this was a problem in recruiting the correct level of labour. Let me explain that in more detail. Ideally what we required to do this task was competent engineers with all the normal engineering skills, plus a little literary skill, i.e.

A. The ability to read a drawing and translate to an understandable catalogue description
B. A good grasp of manufacturing processes and materials

Naturally, our engineers baulked at what they considered a clerical task, which can in some cases be very mundane and repetitive. Therefore we were faced with the necessity to split the task into "Technical" and "Clerical" functions. For the technical function we used engineers. These employees reviewed all components and assigned them to various family groups and also decided on catalogue format. The clerical people then took over and systematically went through the family groups to allocate the full Cee-Code number and draft basic catalogue information. On reflection this does not appear to be such a problem; in fact engineers in other factories may well not have objected, but in our case it was a problem we had to overcome.

Finally, another general problem we are currently encountering is user education. Here we find we need to act as a salesman, because human nature being what it is, people are naturally resistant to change. We have had to assign major portions of time in providing "on the job" demonstrations of how the Cee-Code helps various departments. Some departments are more receptive than others and, of course, all try to fault the system. Therefore the exercise is not without its benefits, and we have a continual refining progress of our system by the users who after applying the system are finding more ways to use its benefits.

Benefits Accrued

Naturally this will be an on going process in all departments as employees become more familiar with the system and its possibilities. We consider ourselves at the start of the benefit period, in that, until now we have been preparing the tools. Now we are beginning to use them.

With inventory visibility now provided our engineering staff have the means of easily identifying and utilising parts and materials already in existence, instead of creating something new, while at the same time, assuring that product quality and performance will not be compromised.

Therefore design avoidance is one of the major benefits of the system. By selecting an existing design rather than creating a new one, the designer initiates major savings in all areas of engineering and production.

In specific design, the mandatory requirements dictate about 10% of the features, the remaining 90% are arbitrary options for which the designer can be guided into selection with the visibility of existing items.

The actual cost of processing an engineering issue is difficult to establish. At Coles we estimate this to be in the region of £500 for each designed part. Audits show that we create approximately 9000 parts per annum. The Cee-Code system enables our engineers to utilise existing parts, which would have otherwise been lost to them, for 16% of their new requirements. Therefore:

9000 parts per year × 16% prevented × £500 per part
= £720,000 per year

Is the amount which can be claimed to be "saved." However, this is really reallocation of existing resources and how we spend these "wooden dollars" for maximum utilization of resources is management's challenge.

Another means of controlling variety expansion is the specifying of the status of the items on file. This is done to gradually move towards a standardised inventory wherever possible.

P—For Preferred—these are easily available relatively cheap high usage items with a logical sequence to their neighbours. These can be used without restriction and catalogues containing only preferred items are freely available to designers.

N—For Non-Preferred—These are items which are candidates for eventual replacement. The designer must obtain supervisory approval for their use in new design and catalogues containing these items are not freely available to designers.

X—For Obsolete—These are items no longer in stock, or on order, and the designer must obtain Central Control approval for their resurrection, the catalogues containing these items not being freely available to any user departments.

As an Example

The effect of this policy on steel together with a variety reduction exercise at the Sunderland factory can be shown by the following statistics:

Class 1—Steel Raw Materials	Jan. 1977	Mar. 1978	Jan. 1979 (Target)
Number of items	1583	1710	1800
Preferred	679	735	800
Nonpreferred	904	570	200
Obsolete	—	405	800

This variety reduction exercise on steel was a combined operation between the Cee-Code office, designers and production engineers. Prior to the Cee-Code system, no number for raw material was allocated by the designers or production engineers. Consequently, it was not easily possible to identify all the parts being made from a specified material. Since the introduction of the Cee-Code system, production engineers have been identifying on the Planning Master the Cee-Code number of the material used to manufacture the part. Thus we can now identify the number of components per raw material number.

We now ask for a print-out which identifies this material number in numerical sequence and the parts to be made from it. With this we are gradually ridding ourselves of the materials which have a viable preferred economic alternative and are only used to manufacture a small number of parts.

An example of RQT 700 Plate is as follows:

Due to the high cost of this special steel (more than £400/Tonne), a policy of ordering special size plates for individual applications, in order to keep the scrap rate to a minimum is followed. As well as ordering and delivery difficulties, this policy caused a major problem in the steel stockyard, because of the number of different size plates being created. Visibility of all existing plate sizes, logically arranged, together with their respective users, given by the

Cee-Code system, enabled overall examination and the deletion of plates of:

A. Nil stock or order cover
B. Low stock
C. Low useage
D. Plates of similar size to neighbours

and at the same time not effectively increasing the scrap rate. With this exercise the following was obtained:

Number of items (6 to 13 mm thickness)	=	180
Number of items deleted	=	132
Number of new items (for logical stocking)	=	14
Number remaining on file	=	62

i.e. the incredible stock reduction possibility of 65%. However, the implementation of this saving is still in progress and the full benefits are not expected to show until 4th quarter 1978.

Raw Material Requirements

Previous to classification and coding, the forward requirements for raw material was based on past useage. Now that a part number is available, identified by engineering, we are investigating a proposal that the production engineers specify the weight of the material used per part on the planning masters. From a "future requirement of parts" print-out the total weight of all materials required can be automatically calculated, and identified. With this information it is expected to substantially reduce the amount of redundant material and also reduce the shortages which occur.

Group Purchasing

As the Cee-Code catalogues now provide inventory visibility for the four Coles U.K. factories, identified by the same Cee-Code number a proficient policy of group purchasing is possible and is being actively pursued at Sunderland.

A simple example being nyloc type self locking nuts. Without the visibility the variation in price per 100 between the four factories was:

M6: £1.20 to £1.95
M8: £1.63 to £2.26
M10: £2.66 to £5.64

With visibility—suitable action was taken to remove these anomalies. Just a small illustration of what is now possible.

Duplicate Part Numbers

Over the years, due to the various engineering activities at the four Coles locations, the Cee-Code has identified the existence of 2154 part number duplications, i.e. identical parts with more than one part number of which 1349 are of the bought in finished (Class 2) type of part. A policy of standardisation of part number on all documentation is being pursued and currently 1162 have been replaced, with all the obvious benefits.

Keep Plates

Again with the visibility provided by Cee-Code we have done an exercise in variety reduction on Keep Plates.

With Cee-Code this type of component is described by shape, thus:

"Rectangular with two holes of the same diameter whose centre line is parallel to the length—centrally positioned." Now considering a material thickness of 5 to 15 mm inclusive we found:

Number on file	=	50
Stocking level	=	30

By introducing minor design changes on some of the items we have reduced the number of items required to 18, and still enable all requirements to be met.

Number on file	=	50
Deleted for nil stock	=	17
Deleted by minor design change	=	15
Modified for logical standard	=	10
Retained on file—no change	=	8
Now on file	=	18

Giving us an inventory reduction on Keep Plate of 64%.

Hydraulic Hoses

An exercise on hydraulic hoses has recently been completed. At commencement there were 1770 different hoses for ¼" bore to 1" bore inclusive. By stating standard design parameters:

A. Type of ends
B. Logical length increments

a range of standard hoses totalling 750 was produced, which in time will replace the 1770 presently existing. As an example, at our Darlinton factory a new model has just been prototyped. All hydraulic hoses used on this machine were taken from the Cee-Code catalogue, no new hoses being created.

Summarising

These examples are of course only a brief summary of what can be done. Further possibilities exist of course and to give you a more concise picture I have listed here how various company departments can benefit.

These benefits all spring from the main objectives of providing visibility and a logical numbering system which collects all similar items together.

Engineering

A. Ability to avoid new design when an existing design will suffice.
B. Ability to readily identify the most cost effective members of a family of parts and design around them.
C. Ability to compare design intentions with similar designs to achieve consistency of design methods and avoid errors.
D. Reduction in prototype costs from use of existing techniques, sources, etc. for similar parts.
E. Ability, in certain cases, to use one drawing with variable dimensions to cover families of parts.
F. Eventually, ability to make best use of computer draughting.
G. Ability to quickly collect information to deal with queries.
H. Ability to apply value analysis and variety reduction to families of products.

Production Engineering

A. Increased ability to introduce group technology into the Machine Shop, therefore greatly simplifying process planning, reducing work in progress, increasing machine utilisation, cutting lead times and increasing output.
B. Increased ability to make good use of numerically controlled machine tools.
C. Reduced set-up costs due to larger batch sizes with smaller variety.
D. Increase visibility of cost standards and also the ability to determine standards by family thereby enabling best use to be made of production engineering manpower for cost savings.
E. Increased ability for nesting procedures for full utilisation of material.
F. Increased fabrication efficiency by standardising piece parts thus permitting their production in bigger batches with reduced set-up time and better ability to optimise material utilisation.
G. Reduced fixtures by combining for family use and identifying existing fixtures.

Other Departments

A. Reduced computing and clerical costs because of smaller files.
B. Visibility of inconsistent costs.
C. Ability to identify lost items.
D. Ability to utilize surplus stocks.
E. Fewer stores locations and arrange stores in Cee-Code order for optimum picking.
F. Ability to find alternative components.
G. Cheaper buying in larger batches.

H. Arrange forward requirements in Cee-Code order for production and buying.

In conclusion, classification and coding will signficantly contribute to the on-going task of management in the running of a successful company, but let there be no mistake, a business can operate without classification and coding. However a much higher degree of proficiency can only be achieved when a good system is available. Similarly the use of computer facilities as exists at Coles is not mandatory, much can be obtained by say a manual card system. However with these tools and a commitment and enthusiasm from management what can be done is limited only by the imagination of the user.

11.5 CASE HISTORY: SERCK AUDCO VALVES, NEWPORT, SHROPSHIRE, ENGLAND, A MEMBER OF THE SERCK GROUP LTD.*

What the Controller Expects from Classification and Coding†

The Company

Serck Audco Valves manufacture Valves for the Gas, Petroleum and Chemical Industries. The valves are used for controlling fluids in pipelines and are produced in a wide range of sizes and materials.

There are approximately 3500 finished products with many variants using 30,000 individual piece parts and raw materials. The turnover of the Company is in the order of £18 million of which 50% is exported. This output is achieved with a payroll of 1000 people.

In normal circumstances 75% of the incoming orders will be met from finished stock.

The manufacturing facility comprises a foundry, machine shops and assembly shops together with supporting services.

The System

The Serck system of coding and classification was introduced in 1960 and covered classes 0, 1, 2, 3 and 4. The system was redesigned in 1968 with the assistance of the same firm of consultants to include products not previously manufactured and will cater for all types of valves known to the industry today.

A decision was made to adopt a "tailor-made" system because it was felt that this would best serve the needs of the company at that time.

*Classification originally commenced in 1960; installed in 1961.

†Presented by V. S. Woolley, Information Services Manager, Serck Audco Valves.

The major benefit being sought was a reduction in the variety of products and components being used (both bought out and manufactured) and considerable success was achieved in this area.

At the time of its introduction a computerised Production Control System and also the Group Technology method of production was being introduced.

The system brings into an organisation the disciplines in the information system which are essential if a computer system is to succeed.

It also makes available to staff responsible for standardisation the information about the components and materials which make up each product and the family relationship which exists between similar parts.

Why Group Technology?

During investigation into our production problems it became obvious that in our environment (multi product small batch) there were two areas requiring urgent attention, these being Material Scheduling and Capacity Planning.

Whilst it is essential that these plans should be co-ordinated it was felt that one should be the dominant factor and the other made to be sufficiently flexible to keep response time down to a minimum.

It was decided to concentrate on Material Planning and Control, because it was felt that if materials are made available to the Workship in the correct quantities at the right time and programmes once issued allowed to remain firm they would, in most cases, be able to cope adequately. At the same time we have gained some experience in attempting to measure and control our Machine Shops but with most unsatisfactory results.

Because of the complexity of setting up a viable control system for workshop scheduling it was decided to look for the solution on the shop floor as it was felt that the functional layout of machine tools which existed was largely responsible for the chaotic situation and subsequent long throughput times with excessive work in progress.

Accepting the hypothesis that the one man business is usually the most efficient it was decided to embark upon an exercise to set up in the machine shops, small self contained manufacturing units capable of machining families of components, these units later became known as G.T. cells.

These cells would be capable of producing large batches or one offs with short throughput times.

In determining the cell machine tool and labour requirement it is first of all necessary to analyse all the components used in each product and then form these into logical families. As these families emerge a large volume of operational information is required and at this stage it is helpful if a computer model of the projected cell loads can be produced giving different machine tool and labour requirements for various mixes in product demand.

Figure 11.3 Group technology module No. 7.

However, once the physical make up of the cell has been determined it may not be necessary to calculate for each programme the machine hours required on each machine as experience suggests that the number of units (Components) is a satisfactory measure of the load, and as the programmes are issued to the factory in fortnightly cycles, adjustments can be made for any shortfall in previous performance.

[Figure 11.3] is a more complex group (Group 7) covering ½" to 4" steel bodies. This group comprises several flow lines, subgroups or cells arranged round a lead machine. In all documentation each cell is identified by a three digit cell number or polycode: the first digit signifying body, plug, component, etc: the second indicating group: and the third, the lead machine. The cell numbers are important for control purposes and help to ensure that Production Control loads a balance of sizes in units to the various cells in the group. Without this finer division of more complex groups to control cells, capacity loading would not be very effctive and control over delivery would suffer.

Author's Note

Mr. Woolley's paper ends at this point. In a paper published in 1968, results of the project were reported by the then managing director, Cordon M.

Ranson, of Serck Audco Valves, through fiscal year 1966–1967, as follows:

1. Sales were up 32%.
2. Inventory stocks were down by £550,000, or 44%.
3. Inventory/annual sales ratio down from 52% to 22%.
4. Average manufacturing cycle time down 12 weeks, to only 4 weeks.
5. Past-due orders down from 6 weeks' output to less than 1 week's output.
6. Number of employees down from 1000 to 987, in spite of a 32% increase in sales.
7. Average earnings per employee increased by 125.45% in the 5-year period.

Mr. Ranson modestly gives credit to the innovations of classification, coding, and group technology for the healthy growth in the profitability of his firm. From firsthand experience, however, the credit for sound decisions

1. To use these management tools, and
2. To plan, organize, and direct the implementation

is due to Mr. Ranson's expertise as a manager.

11.6 CASE STUDY: THE BOEING COMMERCIAL AIRPLANE CO., SEATTLE, WASHINGTON, A DIVISION OF THE BOEING COMPANY*

Classification and Coding in Our Standardisation Programme†

Abstract

This paper describes the classification systems currently in use or under development in B.C.A.C., namely BUCCS-1 Raw Materials, BUCCS-2 Purchased Items, and BUCCS-3 Boeing Designed Piece Parts. Their role in standardisation and the ability to interrelate these systems with a "decision handler" software package to generate standard process plans is described.

The systems described herein are copyrighted by The Boeing Commercial Airplane Company.

*Classification commenced in 1969. Completed after four assignments in 1979, with a 4-year hiatus.

†Presented by G. Millar, Manager BUCCS, The Boeing Company, USA.

This paper gives an overview of the role and status of classification and coding within the Boeing Commercial Airplane Company. Much of what is being done is still in the development stages. However, we have been successful in demonstrating the value and, indeed, the necessity of implementing and *using* these systems to achieve effective standardisation in the engineering and manufacturing disciplines.

To fully appreciate the potential value of classification systems to Boeing it is important to understand a few things about our product line which are not particularly obvious when you fly in them.

I am sure you are all familiar with the four basic models currently in production—the 707 which is still going strong, the 727 our most successful plane, the 737 and the 747.

Each of these has its own series of derivatives—the 747, for example, has eight versions of which the 747SP is only one. Each version has major structural differences from the others.

There are some 175 airlines currently flying Boeing planes—an impressive business statistic but not without some inherent problems for the manufacturing side of the house. The difference between one airplane and another goes much further than the paint job on the outside.

Each airline has its own interior configuration—not just the color of the seats and carpets or the designs on the wall panels—but major hardware differences such as galley location and configuration and flight deck instrumentation. Add to this the one-of-a-kind executive and VIP versions, such as we built for the Shah of Iran and the King of Saudi Arabia and you will perhaps begin to appreciate our "problem."

Under the circumstances it clearly would be difficult to avoid having the same part designed twice unbeknownst to the designers, then to make matters worse by fabricating them differently. Buyers in different divisions are buying the same materials and commodities at different prices and engineers are compounding the problem by proliferating the variety simply because they do not have an accurate or convenient picture of what already exists in the company. These fears were confirmed by investigation and analysis—clearly there was a need for greater visibility. There did not appear to be any practical way of achieving this except through classification and coding! How else could we make order out of chaos; sort ot the good from the bad?

The Boeing Classification Systems group was formed in 1973 and by late 1974 had developed and implemented the Boeing Designed Piece Part Classification which bears the acronym BUCCS-3. The acceptance and success of this system has led to the group being expanded, in 1977, to develop the BUCCS-1 Raw Materials, BUCCS-2 Purchased items, BUCCS-4 Boeing Designed Assemblies, and BUCCS-6 Machine Tools/Process Capabilities classification.

BUCCS	Classifying	Status
1	Raw materials	Implementation
2	Purchased items	Under development
3	Designed piece parts	Fully implemented
4	Assemblies	Not started
5	Machine tools	Under development

Getting these systems developed and implemented in a timely fashion is now of major significance since BCAC will be introducing three new airplanes in the near future:

The 757—a twin engine airplane with a 727 body, new high-technology wing and advanced power plants

The 767—a totally new twin engine, medium range airplane seating 180–210 passengers

The 777—a long range, tri jet seating 200–210 passengers

The development and implementation of these systems have been planned and scheduled to support these new programs since clearly it is here, with some 100,000 part numbers to be released in the next two or three years, that Classification Systems and their applications have the greatest potential benefit for the Company.

Boeing Classification Systems utilise a hierarchical characteristics classification structure and identify the family groupings with a five-digit monocode. Systems requiring specific item identification employ a three-digit, non-significant suffix appended to the monocode.

Boeing Designed Piece Part Classification (BUCCS-3) Structure

BUCCS-3, the first of these systems to be developed and the only one fully implemented at this time, currently contains about 150,000 parts distributed across approximately 5,000 part families.

The nature and grouping of these parts is shown on a graphic form of the Primary Code sheet which represents the upper levels of the hierarchical structure (page 15). Initial breakdown is by shape or function with clearly defined parameters to support the first law of classification, "A place for everything—but only one place."

The parts count overlay on page 16 shows that the "shape" classification contains 80% of the total, breaking down into "axial," "flat" and "bent," with the latter containing over fifty percent of the whole. In fact, twenty percent of the parts are single bend, bracket-type configurations—certainly a good area to encourage engineering and manufacturing standardisation since there would be little conflict with the cause of technological advancement. A bracket is still a bracket!

Procedure

The flow diagram shows that the engineer or draftsman makes use of BUCCS-3 by taking his preliminary sketches or layouts to the BUCCS file located in the area. There a BUCCS coder codes the sketch and pulls out a folder bearing the same code. The folder contains about twenty 8½ × 11 inch mono-detail drawings for the engineer to peruse and hopefully find a part which will suit his purpose. He will then call out that part number on his parts list. Even if he does not find a part that exactly fits his needs he may find one that can be used with minor modification. At the very least, he can gain some useful data from studying designs similar to the one he is about to create.

The cumulative success rate in BCAC for retrieval of designs and significant design data is ten percent, which is considered to be good in an industry which combines high technology with extreme weight consciousness.

Another highly valuable service is made available to the designer at this time. The folder of drawings also contains a series of "Producibility Tip Sheets" pertinent to the subject part family. These "Tips" recommend design practices which can realise significant economies in fabrication of the part and play an important role in providing a much needed communication link between manufacturing and engineering when and where it is needed—at the design stage "Tips" help to create part designs that recognise current fabrication requirements and identify the least costly "design for manufacturing" when that engineering option is possible. If he does not find a "reusable" part the draftsman completes his design and includes the BUCCS-3 code on the drawing. The drawing then goes back to the BUCCS coder for code verification and buy-off prior to being released. As part of the normal release routine a microfilm copy of the new drawing is sent to Classification Systems so that it can be added to the BUCCS-3 files. Putting the BUCCS-3 code on the drawing enables Manufacturing to use the system without the need for a qualified coder. Using the code to access a file identical to that used by engineering the planner can retrieve existing processing and the tool designer can retrieve and adapt existing tool designs and even modify existing tooling for multiple part use. Group Technology applications of the code have led to better order groupings and equipment utilisation.

"Standards" Applications

The biggest problem with a Standards System is not deciding what to put into it but what to keep out of it. How to control growth without hampering the designer's ability to advance the state-of-the-art, particularly in the Aerospace industry where weight comes second only to reliability. Standards Systems without objective and realistic controls can easily grow until they are so cumbersome that it becomes easier to "design from scratch" rather than take the time to search for a standard.

This could soon become the case in BCAC if we do not find a way to exert some control. One way is through the visibility afforded by the contents of the BUCCS-3 system. By objectively comparing the contents of high volume families containing standard-like parts with the corresponding standards we have been able to recommend reductions in those standards.

A good example of this is the standard for simple, right angle brackets which allows over half a million combinations of gauge, length, flange width and hole quantity. Statistical analysis of the twenty four thousand parts in the corresponding BUCCS-3 families can give us the visibility to greatly reduce this number.

Raw Materials (BUCCS-1) and Purchased Items (BUCCS-2) Classifications Background

Development of BUCCS-1 and BUCCS-2 was started in April 1977. BUCCS-1 is currently being implemented while implementation of BUCCS-2 is scheduled to be completed by the end of 1978.

The need for these systems was emphasised by some significant occurrences experienced at the BUCCS-3 files. Draftsmen and engineers would frequently search the files for items which the BUCCS coder would advise them were "Standards," to be found in the Boeing Standards Manuals. The problem is that the Standards System has grown over the years with little real discipline until it is too large to be used conveniently. The user often "cannot find what he is looking for," even if he knows it should be there! During our initial developmental analysis we discovered many examples of the kinds of problems facing the user of the Standards System. Items which should have been grouped together were often separated by nomenclature. An example of this problem is the J-section extrusions which are to be found under the heading of "angles" while other J's are to be found under "channels."

Clearly we have a retrieval problem and classification would do much to solve it, but the real question is—do we need all those standards? How do we know when we have enough standards? How do we know when we have too many? To try and answer these questions we elected to use as our database for BUCCS-1 and BUCCS-2 those materials and commodities which are currently in use or are being purchased for use. The Material Organisation was able to provide us with pertinent data on all such Boeing Standards and vendor items.

By restricting ourselves to in-use items, we are taking a first step towards realistic "standards." A standard item should be a PREFERRED item. The contents of BUCCS-1 and BUCCS-2 are "preferred" only by virtue of their having been previously used.

The next step is to further analyse the database to screen out the unjustifiable duplicates. This is a task which requires authoritative, coordinated direction from Engineering, Manufacturing, Material, Materials Technology and the Standards Organisations. BUCCS-1 and BUCCS-2 will give them the visibility to perform this task.

Structure

Both systems employ hierarchical structures similar to BUCCS-3 to group contents to the "family level", identified by a five-digit monocode (1XXXX and 2XXXX). Each family is then displayed in tabular format giving those pertinent characteristics of the member items necessary for specific identification by the user. A randomly assigned, non-significant, three-digit suffix to the code identifies each item. Currently there are about seventeen thousand materials and material configurations in BUCCS-1 and forty-eight thousand items in BUCCS-2.

Procedure

It is intended that the user will access the system directly without the assistance of a BUCCS coder as has been found necessary with BUCCS-3. It has been our experience that many users encounter difficulties when using matrix type code sheets. It takes time to become familiar with them and there is high risk of error by the infrequent user. We have overcome this problem by using an alphabetically sequenced keyword index to derive the family monocode.

The user then enters the five-digit monocode on a computer terminal to call up a display of the "catalogue page" on the CRT screen. He can then select a material or part to suit his needs with a high assurance that it is indeed available and there is little or no risk of rejection downstream resulting in delays and the added costs to change the drawing. The complete eight-digit code is added to the item call out on the parts list and will be used by Material and Manufacturing as a means of more economic purchasing practices and better shop order grouping.

Applications: Data Retrieval

Most of the time these systems will contain enough information for the designer to make his selection. To include every piece of available information about each item would be impractical, if not impossible. However, it should be borne in mind that classification and coding is, first and foremost, an indexing technique. Thus data files of test, performance, design allowables, and other pertinent data can be constructed and accessed using the BUCCS code. We are planning to use this approach as the basis for a system which will generate a "Certification Report" specific to each new airplane.

Improved Standards

As in the case with BUCCS-3, these systems will provide the main ingredient for the objective identification of standards—visibility of what is actually used!

This is particularly true of BUCCS-1 and BUCCS-2 since seventy-five percent of their contents are Boeing Standard Items yet the actual quantity represents only a fraction of the contents of the current Standards System.

Boeing Generative Process Planning System

Background

Optimum standardisation and automation of Process Planning has long been a goal of the Boeing Company. The recently implemented On-Line Planning System computerises and automates much of the procedure for creating a plan. However, the planner is still left with the often subjective decision-making relative to the particular process and equipment to be used. Experienced and skilled, as most of our planners are, there is nothing in the present system to prevent them on different occasions from selecting different processes, possibly varying widely in cost, for the same type of part made from the same material.

The theoretical solution is fairly obvious:

> Parts with identical or nearly identical design characteristics and requirements should be manufactured using the same processes.

Devising the method to do this in a practical and timely fashion was not so obvious. However, by early 1977 we had gone as far as recognising that most of the design characteristics of a piece part pertinent to manufacturing the part were contained in the BUCCS-1 and BUCCS-3 codes. Our problem was to relate the coded part description to manufacturing equipment and process capabilities with manufacturing decision logic.

The answer came in June 1977 when a western (America) University's Manufacturing Technology Department offered a software package that had been developed to function as a decision handler for general taxonomies (classifications) with applications in manufacturing.

A task team was put together in the summer of 1977 to demonstrate the feasibility of generative process planning utilising Boeing classification structures (BUCCS-1 and BUCCS-3) and the BYU decision handler (DCLASS).

Scope

The generative planning system model which was developed, was conceived as a potential enhancement to existing computerised process planning systems. Its purpose was to demonstrate that uniform manufacturing process plans could be generated from design information. Sheet metal piece parts fabricated in channel form were selected for the feasibility demonstration. This type of part is common in airframe structures (comprising 7% of the total designed piece part count) and the fabrication processes are of moderate complexity, averaging about 12-16 operations per part. It was assumed that if process planning of this complexity were generated, then the general applicability of the concept would be apparent.

System Description

The system concept involves the interrelations of several logic elements and a text file within a software package to form a truly unique *generative* planning system. The total system consists of:

1. Classification logic for part shape (BUCCS-3) and raw material (BUCCS-1).
2. Special process parameters.
3. Manufacturing decision logic that relates drawing derived shape, material and special characteristics to manufacturing equipment and process capabilities.
4. An operations narrative file which describes each potential manufacturing operation that the factory can perform with the available manufacturing equipment and processes.
5. Sequencing decision logic which arranges the selected operations in the proper order.
6. A plan preparation segment to output a process plan in the desired format.

The logic elements and text file are interrelated through computer sensible internal codes that identify their inter-relationships.

Conclusion

This method for generating process plans from engineering data holds great promise. It is being evaluated at the Boeing Commercial Airplane Company as an included enhancement to the process planning portion of the overall operations management systems. Generative process planning exhibits a high potential for accomplishing several universally accepted goals for automated process planning: uniformity, consistency, and optimisation—the key elements of standardisation!

11.7 CASE STUDY: AN OVERVIEW*

A Top Management View of Classification, Coding, Standardisation and Group Technology †

In presenting this paper I am dealing with two subjects. Firstly, there is the subject of Classification and Coding. Secondly, there is the separate but interrelated subject of Group Technology.

*Classification commenced in 1961, installing five assignments over a 4-year span.

†Presented by W. Jack, Managing Director, J.L.G. Ltd.

A. *STANDARDISATION AND SIMPLIFICATION*

1. The subject of Classification and Coding in my experience has ramifications and possibilities as little understood as those of the computer 15 or so years ago.
2. Classification and Coding is a basis for standardisation programmes. Top management can establish control in a great number of ways in many different departments. Classification and Coding can be the means for saving considerable sums of money through variety Control and variety Reduction.
3. I have been asked to tell you of my experience over a number of years in the field of Classification and Coding and in the field of Group Technology.
4. In 1961 I had a problem with regard to a tools organisation and after survey the Classification of our general purpose perishable tools revealed that 32.3% were non-standard.
5. The actual cost in the five years after Coding showed a reduction of 30% per annum.
6. In the area of Special Purpose designed tools the estimated reductions in tools designed over a period of 6 years was 33.9%.
7. In examining tool costs as a percentage of sales revenue, a reduction was achieved—from 2.64% to .81% of sales.
8. With regard to labour costs, a reduction in number of people employed in the tool design and manufacture activity was achieved—from 106 to 82, i.e. 22.6%.
9. It soon became apparent that much could be achieved through development of Classification and Coding in primary materials. After Classification and Coding it was revealed that 9.9% were Non-Preferred, 4.5% to be deleted and 4.5% were duplicates, a total reduction in number of items of 18.9%.
10. Purchased commodities were then classified and coded and a reduction of just over 7% was achieved at the outset through the elimination of duplicates. A further 27.2% were classed as Non-Preferred—a total reduction of 34.2% of pre-coding variety.
11. Having achieved significant savings in tools, primary materials and purchase commodities we examined our design components and submitted for Classification purposes 29,492 designs. Prior to Coding we eliminated 9%, found that 5% were obsolete and a further 5% were duplicated. The *initial* total reduction of designs on file was 18.2%.
12. In the *5 years after Coding* we estimated that our reductions in numbers of single piece part drawings was of the order of 49.8%.
13. A re-check of figures in an attempt to find out if the 49.8% reduc-

tion was valid, revealed that although there was a reduction in activity after Coding, an improvement of 41.8% could still be documented.

14. *Acquisition and Integration*
    Company A took over Company B and an extension of the Classification and Coding System to Company B resulted in worthwhile benefits. Company B had 8% Duplicates in Primary Materials whilst 7% of the items were common to both companies.
    a. In the field of Purchased Commodities, Company B had 17% duplicates whilst a further 7% were found to be common to both companies.
    b. Largely, what I have said up to the present about the acquisition has involved the establishment of control without necessarily achieving savings through variety reduction exercises. However, variety reduction efforts in one small area covering 2231 items of purchased commodities, a reduction of 28% was achieved.
    c. Furthermore, over a period of 4 years of further effort net reductions were achieved in the field of purchased commodities equal to 13% of total catalogued items.
15. The cost of creating a new part varies from company to company. Nevertheless, the savings achieved by not creating ***unnecessary new designs*** must be considerable.
16. The Classification and Coding programme was not just something that happened or which affected one function of the business. It affected the purchasing organisation, stock control, methods, progress, the tooling organisation, planning, time standards, data processing and cost accounting. It cut across departmental barriers in much the same way as happens when one installs a computer. Barriers exist departmentally in many companies but we found that with Classification and Coding, administration personnel became enthusiastic because their work was much simplified. It was particularly because of this, that we found we could project Classification and Coding beyond the administration functions and into methods of manufacture.

B. *GROUP TECHNOLOGY*

1. We had a serious problem with very small lots of parts. The usual cycle from receipt of order to completion exceeded 12 weeks. The shops were reluctant to run the small lot order as they were on a wage incentive scheme.

   The majority of these small lot parts were for spares and service.

So in 1963, we decided that perhaps Group Technology might hold a solution for this problem. We had first heard of the technique at a London Seminar conducted by the Consulting specialists who had installed our classification and coding scheme. We decided to investigate Group Technology potential first. We worked together to plan and implement the first G.T. unit as a pilot study to determine:

a. If in fact it would help us solve our small lot problem and
b. What other benefits could be gained in improved costs elsewhere in the shop on *not* so small production lots.

2. We examined the order on hand and found that the majority of candidate parts were in sub-class 34000, round axial parts. In addition to geometry, we then established a number of additional parameters to qualify a part as a candidate to run on the G. T. unit—as yet not established.

a. Lot size must be 12 or fewer on the order.
b. Only *steel* parts were to be considered so as not to mix swarf requiring clean-up between lots.
c. We set size parameters, the limit of which was the largest practical diameter that a three jaw universal chuck would hold, and a length that would not require a steady rest.

We then developed composite components. This was done by analysing the most common shapes that met the three parameters and included all the features that might be found on the parts.

We then processed the composites of which there were two, and selected the type of equipment.

3. A word about the equipment. Because this was a laboratory trial set up (pilot scheme), we used old, general purpose machine tools that had long since been amortized. The capstan and turret lathes were more than 15 years old (one was 20 years of age). The 2 Cincinnati mill was 18 years old and the Cincinnati Carlton radial drill was of the same vintage.

We fixed these up a bit—tightened the gibs, replaced a head stock bearing or two, but nothing major was done to rebuild them.

4. Then we established permanent tooling stations for each of the two turning machines. We selected the accessories and tool holders to produce the shape configurations and features on the composite components. The capstan lathe was set with new colletts, the turret lathe with a rather expensive 3 jaw universal chuck capable of holding run out to two tenths (.0002") total indicator runout.

We equipped the radial drill with and index table with accuracies in the ± 5 second range for drilling on bolt circles, mainly.

5. The parts selected were produced in an average in-process cycle of 4 hours from receipt of material to finished part. When compared to the 12 + weeks prior to G. T., this certainly meant we had solved the short order service and spare parts problem.
6. We kept very accurate costs and found that set-up costs were reduced by more than 80%, and each piece costs were lower by 15%—in spite of working under a day work (non-incentive) scheme, while the rest of the shop were on incentive.
7. We added machinery to the group to enlarge the scope of parts which could be accepted in the unit ending with 7 machines and 4 operators.
8. A second G. T. unit was also established of a comparable size to run other composite parts and parts with somewhat different processes.
9. There were problems. The old machinery broke down. The utilisation run versus idle was up to 85% and more, and the old machines couldn't stand the load. So we replaced them with plain, unsophisticated machine tools. There was not one N. C. Machine in the lot.
10. During the selection of parts for setting the specific tooling, we discovered anomalies such as variations in fillet radii, chamfers, thread lead angles and the like. We remedied these by a more encompassing set of drafting standard practices and the establishment of consistent shop practices.
11. Our G. T. pilot made a dramatic impact upon cost, delivery and quality. Oh yes, I forgot to mention that scrap and reoperation costs were well below the shop average.
12. Our major problems were few. One was that the unit ran itself out of work. From a backlog of 12 + weeks on hand when we started, the backlog disappeared within three months, and operates on less than one day's backlog. This also caused a second problem, increasing the lot size limitations. Even after they were raised to 30 or fewer (from 12 initially), planners tried to sneak in larger orders for speedier delivery.
13. The conventional cycle at 1½ to three days per operation in the process layout of the shop (by machine function) and with 7.2 operations per part, the cycle is 3 weeks instead of less than one day in the G. T. units.
14. One benefit was a vast reduction in the paperwork. From 11 documents and inspection sign-offs in the main shop to just 3 in the G. T. unit.

Overall, we were most gratified to have had our first experience with G. T. turn out better than we expected in all respects.

And we were also gratified to have been first to do this in the U.K.—not because we were first so much, as to provide an example to others to help our industries to be more competitive. It is in this latter that I take most pride and satisfaction.

C. *SUMMATION*

1. In summary, some of the benefits achieved from the classification and coding and Group Technology programme were as follows:

a. Initially on Coding	Reduction
Primary materials	– 10%
Purchased commodities	– 7.3%
Design components	– 10%
General purpose tools	– 3%

b. After Coding	Reduction
Special purpose tools	– 33.9%
General purpose tools	– 30%
Employees (tools)	– 22.6%
Design components	– 41.8%
Purchased commodities	– 28%

c. Group Technology
Unit effective time up from 60% to 74%
Set up time down by 80%
Each piece time down by 15%

2. I have mentioned that the departmental barriers are reduced or elminated in the programme as I have outlined.
3. In my experience, top management is not aware of the value or potential of Classification and Coding or of Group Technology. I believe many middle managers are aware of the existence but have not the authority to get on with the job.

D. *PERSONAL CONCLUSIONS AND OBSERVATIONS*

1. Classification and Coding linked with the computer facility will give an opportunity for total management involvement and will give management a tool of considerable power.
2. I believe that a good Coding system is a pre-requisite for good Group Technology. I am convinced that whilst you can get returns from good use of a computer, the returns in the end will be in-

creased tremendously if a good Classification and Coding system is in existence.

Likewise the returns from a good attempt at Group Technology will be enhanced if a good Coding System is in operation.

3. What has happened to Group Technology in the U.K.? In the late Sixties it got into the hands of Government sponsorship—then in came the academics. Both groups failed—then in came the generalist management consultants who also failed.

   Group Technology failed in the U.K. because it was thoroughly misunderstood at the outset. It was defined by me in 1966 as "The grouping of parts in such a way that the tooling, method planning and production of each part are not considered in isolation but in the context of a whole group of similar parts."

   In order to do this one must have a system for proper retrieval of information.

   This pre-requisite was ignored in the U.K. because of the poor interface which we have between design and production functions.

4. In the U.K., Engineering education emphasis is on research and science whereas in the U.S.A. it is on practical application.

   Educational leaders in the U.K. in Polytechnics and Universities are mostly scientifically minded with no experience or little regard for the practical requirements of industry.

   Cost improvement engineering and real industrial engineering receive no attention from scientific professors who are much concerned with higher learning.

5. For Group Technology to succeed I believe you require beforehand a good Coding System and an understanding that a Group Technology Unit application should be alongside functional layout of machine tools and not in place of it.

11.8 CONCLUSIONS

The use of classification, coding, and group technology has been clearly demonstrated to be of considerable value to the firms whose cases have been presented. Quantified monetary benefits overwhelmingly support the efficacy of their decision for embarking upon programs to use these management tools.

But what of the nonquantifiable benefits in the cases reported? What, for example, is it worth to spare managers the need to become involved in decisions that can now be made at an operating level? The benefits can easily outstrip all of the quantified savings reported.

When Mr. Bunting alluded to "wooden dollars" and how to organize the freed resources for their optimized use as a challenge to management, he is saying that there was time to think, to plan, to organize, and to execute. It is true that when relevant facts are available to managers, the decision-making process requires less time, while improving decision quality.

Also, freeing managers from tactical problem solutions allows additional time for strategic planning decisions for the future. To our way of thinking, these benefits multiply the quantified benefits by an order of magnitude, at the very least. For this reason alone, no firm, institution, or agency can ignore an opportunity to increase the productive effectiveness of its managers. To cope with the future and still make a profit today, therefore, requires managers to decide upon only those tactical problems that are exceptions. This is the greatest benefit of all and the most compelling reason for adopting a classified and coded data base. The improvement of our management productivity is imperative.

Postscript

Conclusions and Recommendations

The primary postulation and recurrent theme in the preceding 11 chapters has been that you cannot manage without visible, relevant data. Repeatedly, we stressed the need to establish a discipline for controlling data that is fail-safe, foolproof, and mutually beneficial to all parties. We consistently sought to prove that only by classifying and coding, simplifying, and standardizing a given population of data is it possible to eliminate redundant effort, raise productivity, and lower costs. This was done within the context of an employer/employee relationship.

In this postscript, the stage setting and roles are different. Now, it is supplier and consumer. The business? Communication! The suppliers—for simplicity—are two corporations whose services are widely used by all of us virtually every day of our lives. We are the consumers.

Both firms have spent billions of dollars to update equipment, improve methods, and design new systems and new physical plants. Both firms have created systems for providing their services that are dependent upon the consumers' use of standardized codes and procedures built upon these coding systems.

But this is where the similarity ends. One firm prospers, the other does not. The divergent results can be traced to the success of the one firm (company A) to enlist the consumer to use the system as designed and the virtual failure of the other (company B) to accomplish this end.

Company A, in its system of servicing communication needs, has done the following:

1. Made codes mandatory to complete the transaction, irrespective of which of several options available to the consumer are used
2. Provided tutorial assistance that enables the consumer to receive the proper code easily, quickly, and without cost
3. Provided a positive incentive to use system codes with self service, rewarding the user with much lower cost of service (in fact, all service is less costly, irrespective of option)
4. Errors are forgiven and no charge is made for them.

Company B, in servicing its customers, has done the following:

1. Made no requirement that the codes be used to complete the transaction.
2. Provided no positive incentive in the form of lower costs, if the codes are used, threatening only poorer service if the codes are not used. Service is uniformly poorer and more costly for all.
3. Errors are at the users' expense.

One organization is Bell and the other telephone systems. The other organization is the United States Postal Service. You decide which is company A and which is company B.

The conclusion is obvious. Without built-in discipline and incentives, even the best system ever developed will fail.

The technology described in this book is exploding. The evolution rate over the past 31 years of our firm's experience has speeded up by an order of magnitude in the most recent 10-year period.

As our nation's productivity improvement rate continues to decline, bold—perhaps even risky—decisions have to be made, and action taken as quickly as possible, if the condition is to be remedied.

The first requisite of management, in its prime role of decision making, is to get the facts. Facts are data. Therefore, to get the facts means having the capability to retrieve them and make them visible.

The classification, codification, simplification, and standardization of data, therefore, are fundamental imperatives. It is earnestly hoped that this book will help to shed a little light on some ways and means to make this happen.

In what order of priority this is best done requires careful study.

To create visibility of the manifold and complex interrelated activities and items with which we daily deal is plain, common, no-nonsense horse sense.

Putting patches on cancerous growths only hides the problem—it doesn't solve it. We should recall once more the words of Morris L. Cooke when, more than three-quarters of a century ago, he prophesized:

> Only as we learn to classify and to code, to simplify and to standardize, will any real science of management emerge.

We rest our case.

Index

UNIV DUNELM

DURHAM UNIVERSITY LIBRARY
3 0104 01091963 1